D0961799

MORE PRAISE FOR *DEVIATE*

"Combining evolutionary imperatives with modern imaging of the brain, *Deviate* helps us understand perception as the key to an individual's survival. It is written with humor, clarity, and delight. I highly recommend it."

—Jerry Harrison, lead guitarist of the Talking Heads

"It's time to deviate! Citizens of the world seem stuck in their paths, and are losing perception of what's there to enjoy as the mundane traps us in our tracks and routines. Beau Lotto teases our sense of adventure by suggesting we romp through our perceptions and break out of the framework. *Deviate* will give you a sense of yourself, whether you're a misfit or wish you were one!"

—Marian Goodell, co-founder and CEO of Burning Man

"What if we all tried harder to be misunderstood? And what if we could embrace and channel our own misunderstanding of the world around us? Beau Lotto's *Deviate* honors the messy, imperfect genius of human perception as the most valuable resource for creative progress. Lotto is teaching us something so loudly fundamental to our existence, it seems almost impossible that we've missed it."

—Ross Martin, executive vice president of
Marketing Strategy and Engagement at Viacom

"In *Deviate*, Beau Lotto's remarkable research into human perception is crystallized into a series of astute explanations of how we experience reality. By bringing together an 'ecology of the senses' that goes beyond the mechanisms of the eye, Lotto's ingenious account of the brain's perceptive evolution arrives at an extraordinary proposition of how we can go beyond our current ways of seeing. . . . It is a brilliant book!"

—Hans-Ulrich Obrist, director of the Serpentine Galleries and author of *The Interview Project*

"Beau Lotto has delivered a fresh, provocative, stimulating, revealing, neuro-inspired, entertaining text on that most fugitive of subjects—reality. . . . The world of theoretical and experimental neuroscience has much to offer us as we search to produce better environments for all."

—Ian Ritchie, director of Ian Ritchie Architects and architect of the largest free-standing glass building in the world

"Beau Lotto shows better than anyone else how dependent we are upon our own limited sensory perceptions of the world. The radical thesis that he presents in *Deviate* reveals to us that reality is relative, and that we, ultimately, are capable of changing our world through changing our perception of it."

—Oafur Eliasson, sculpture artist and spatial researcher, founder of Studio Olafur Eliasson

"In a brilliant and skillful way, Beau Lotto pulls the rug from under our naive view of reality—bit by bit. In reading this book, we discover how our conventional way of seeing, of perceiving reality, is incomplete and illusory. He begins to dismantle this illusion by showing us why we see the world the way we do and, in doing so, he opens the curtain to a new beginning—a new beginning of seeing past our individual interpretation of reality, to recognize that others may surely have a different interpretation. In daring us to deviate, Lotto encourages us to discover that compassion has a root that can be revealed through scientific insights."

—Peter Baumann, founder of Tangerine Dream

"Beau Lotto is one of the most creative scientists I know, and his passion for introducing neuroscience to the public ranks him among those rare communicators like Carl Sagan whose ideas can change people's thinking. At a time when many neuroscientists are pursuing the mindless goal of mapping all the connections in the human brain, Beau is right on target in his conviction that science advances by doubting the conventional wisdom and asking simple questions in a novel way."

—Dale Purves, professor emeritus at the Duke Institute for Brain Sciences and member of the National Academy of Sciences

"As a neuroscientist and a specialist in vision, Beau Lotto opens up the subject of just how it is possible to actually see and understand anything in the world when it seems that meanings are always constructed somehow separately from the reality of what we see. This is done with immense clarity and ease . . . directly relevant to anyone involved in shaping our world—designers, engineers, and architects."

—Alan Penn, professor of architectural and urban computing at University College London

"If someone else told me that reality is something we create in our heads—I'd up my medication. This brilliantly written book shows us that this is actually the road to liberation. We have the ability to change our internal landscapes, making our lives a masterpiece rather than a 'been there done that' cliché."

—Ruby Wax, OBE, comedian, actress, mental health campaigner, and bestselling author of *How Do You Want Me?*

DEVIATE

DEVIATE

THE SCIENCE OF SEEING DIFFERENTLY

BEAU LOTTO

Illustrations by Luna Margherita Cardilli and Ljudmilla Socci

 hachette
BOOKS

NEW YORK   BOSTON

Hachette Books
Hachette Book Group
1290 Avenue of the Americas
New York, NY 10104
hachettebookgroup.com
twitter.com/hachettebooks

First Edition: April 2017

Hachette Books is a division of Hachette Book Group, Inc.
The Hachette Books name and logo are trademarks of Hachette Book Group, Inc.

The publisher is not responsible for websites (or their content) that are not owned by the publisher.

The Hachette Speakers Bureau provides a wide range of authors for speaking events. To find out more, go to www.hachettespeakersbureau.com or call (866) 376-6591.

Interior Design © Luna Margherita Cardilli and Ljudmilla Socci, Black Fish Tank Ltd, 2017

Library of Congress Control Number: 2016962966

ISBNs: 978-0-316-30019-3 (hardcover); 978-0-316-30017-9 (ebook)

Printed in the United States of America

LSC-C

10 9 8 7 6 5 4 3 2 1

To perceive freely . . .
Through tempest . . .
Violence un-cast . . .
With courageous doubt . . .
A tilted self . . .

Dedicated to those who walk tilted.

CONTENTS

ACKNOWLEDGMENTS

All knowing begins with a question. And a *question* begins with a *"quest"* (to state the obvious), as does life. At the core of living, then, is the courage to move, to step with doubt but step nonetheless (sometimes off a cliff, which is a less good step). Fortunately, no one steps alone (except that last one). My shuffles expressed here were and are enabled by the courage of others who in different ways enable me to live: My deviant Mum and Padre and Janet, my four mad sisters, my gorgeous gremlins Zanna, Misha and Theo, and my essential and beautiful (co-)explorer and creator Isabel. All incredibly colorful people who have shown me new ways of seeing, sometimes against my will (sorry), always to my benefit eventually. They are *"my why,"* my foundation for attempting to see freely, and the motivation to support others in their attempt to do so.

I thank my teachers (and all teachers more generally). Most of our life happens without us there, since most of our perceptions were seeded by, if not out-right inherited from, others. Of particular importance to me have been the perceptions of one of the world's leading neuroscientists Dale Purves, who was the initiator and distiller of my way of thinking and being in science and the science of perception. A mentor in the truest sense. Dale, along with Richard Gregory, Marian Diamond, Joseph Campbell, Houston Smith and Carl Sagan and their deviating-ilk reveal in action that *true* science (and creatively-critical-thinking in general) is *a way of being* that can transform. They are teachers who show us how to look (not what to look at). Teachers like Mrs Stuber at Interlake, Mrs Kinigle-Wiggle and Marshmellow, Mr Groom and Orlando at Cherry Crest, thank you. I also thank my core collaborators (teachers of a different kind): Isabel Behncke, who has expanded, opened and grounded my knowledge personally and academically in essential ways (including the different grounds of Chile the kelp-beds to lake-beds), Rich Clarke, who has been core to the lab's activities and ideas since its inception, Lars Chittka, who taught me how to train bees, Dave Strudwick, who was essential to creating the lab's science education programme . . . and my diversity of PhD and Masters students in neuroscience, computer science, design, architecture, theatre, installation art, and music, such as David Maulkin, Daniel Hulme, Udi Schlessinger and Ilias Berstrom, who became experts in areas that I was not, and in doing so complexified the lab and my thinking in essential ways.

I also thank my highly engaged editors Mauro, Bea and Paul, my brilliant agent and friend Doug Abrams (whose ambition and impact in publishing is inspiring), and my tremendous support-writer Aaron Shulman without whom this 20-year project would never have been seen by me, much less anyone else. Together we struggled to innovate; i.e., to balance creativity and efficiency (or more accurately, they struggled to balance me patiently).

And I thank *you*. One of the most challenging things we can do is to step into uncertainty. I conceived of *Deviate* as an experiment in book-form, a place to share my necessarily limited understanding of perception as well as my speculations and opinions (inherited and seeded) in the hope—and I can only hope—that you would know less at the end than you think you know now, and in doing so understand more. In nature form (or change) comes from failure, not success. The brain—like life—does not search to live, but to not die. Which makes success an accident of what failure leaves behind when one is thoughtfully deluded enough to walk tilted (long) enough.

The only true voyage of discovery . . .
[would be] to possess other eyes,
to behold the universe through the eyes of another.
—Marcel Proust

The Lab of Misfits

When you open your eyes, do you see the world as it really is? Do we see reality?

Humans have been asking themselves this question for thousands of years. From the shadows on the wall of Plato's cave in *The Republic* to Morpheus offering Neo the red pill or the blue bill in *The Matrix*, the notion that what we see might not be what is truly there has troubled and tantalized us. In the eighteenth century, the philosopher Immanuel Kant argued that we can never have access to the *Ding an sich*, the unfiltered "thing-in-itself" of objective reality. Great minds of history have taken up this perplexing question again and again. They all had theories, but now neuroscience has an answer.

The answer is that we don't see reality.

The world exists. It's just that we don't see it. We do not experience the world as it is *because our brain didn't evolve to do so*. It's a paradox of sorts: Your brain gives you the impression that your perceptions are objectively real, yet the sensory processes that make perception possible actually separate you from ever accessing that reality directly. Our five senses are like a keyboard to a computer—they provide the means for information from the world to get in, but they have very little to do with what is then experienced in perception. They are in essence just mechanical media, and so play only a limited role in what we perceive. In fact, in terms of the sheer number of neural connections, just 10 percent of the information our brains use to see comes from our eyes. The rest comes from other parts of our brains, and this other 90 percent is in large part what this book is about. Perception derives not just from our five senses but from our brain's seemingly infinitely sophisticated network that makes sense of all the incoming information. Using perceptual neuroscience—but not *only* neuroscience—we will see why we don't perceive reality, then explore why this can lead to creativity and innovation at work, in love, at home, or at play. I've written the book to *be* what it describes: a manifestation of the process of seeing differently.

But first, why does any of this really matter to you? Why might you need to deviate from the way you currently perceive? After all, it *feels* like we see reality accurately . . . at least most of the time. Clearly our brain's model of perception has served our species well, allowing us to successfully navigate the world and its ever-shifting complexity, from our days as hunter-gatherers on the savannah to our current existence paying bills on our smartphones. We're able to find food and shelter, hold down a job, and build meaningful relationships. We have built cities, launched astronauts into space, and created the Internet. We must be doing something right, so . . . who cares that we don't see reality?

Who cares that we don't see reality?

Perception matters because it underpins everything we think, know, and believe—our hopes and dreams, the clothes we wear, the professions we choose, the thoughts we have, and the people whom we trust . . . and don't trust. Perception is the taste of an apple, the smell of the ocean, the enchantment of spring, the glorious noise of the city, the feeling of love, and even conversations about the impossibility of love. Our sense of self, our most essential way of understanding existence, begins and ends with perception. The death that we all fear is less the death of the body and more the death of perception, as many of us would be quite happy to know that after "bodily death" our ability to engage in perception of the world around us continued. This is because perception is what allows us to experience life itself . . . indeed to see it as alive. Yet most of us don't know how or why perceptions work, or how or why our brain evolved to perceive the way it does. This is why the implications of the way the human brain evolved to perceive are both profound *and* deeply personal.

Our brain is a physical embodiment of our ancestors' perceptual reflexes shaped through the process of *natural selection*, combined with our own reflexes as well as those of our culture in which we are embedded. These in turn have been influenced by the mechanisms of development and learning, which results in seeing only what helped us to survive in the past—and nothing else. We carry all of this *empirical* history with us and project it out into the world around us. All of our forebears' good survival choices exist within us, as do our own (the mechanisms and strategies that would have led to bad perceptions are selected out, a process that continues to this day, every day).

Yet if the brain is a manifestation of our history, how is it ever possible to step outside the past in order to live and create differently in the future? Fortunately, the neuroscience of perception—and indeed evolution itself—offers us a solution. The answer is essential because it will lead to future innovations in thought and behavior in all aspects of our lives, from love to learning. What is the next greatest innovation?

It's not a technology.
It's a way of seeing.

Humans have the wild mpdlkkt thas fbr dotig ke ednxoa oe ds wis
and generative gift of ioj yudrtfgyu jokmplifefbsv98 igharogf
being able to see their lives ari uir albwireuty owuegoiw sydpo
and affect them just by reflecting xad v aufdagv ieudagfx oqpie
on the process of perception itself. We swii ikglisdykghw lisdw
can see ourselves see. That is what this 8 aada gsdf voilak

wsunta sydi book is fundamentally about: seeing your see or per-
ukfyhv ows ceiving your perception, which is arguably the most
fdkg fcof essential step in seeing differently. By becoming aware of
airk hq the principles by which your perceptual brain works, you
zu oke can become an active participant in your own perceptions and
in this way change them in the future.

Down the Rabbit Hole

Alice follows a white rabbit down a hole and ends up in a world
in which fantastical things happen. She grows in size; time is
eternally stopped for the Mad Hatter at 6 p.m.; the Cheshire Cat's
grin floats in the air, sans the cat. Alice must navigate this bizarre
new environment and at the same time maintain her sense of
self, no easy task for anyone, let alone a child. The book *Alice
in Wonderland* underscores the virtue of being adaptive when
confronting shifting circumstances. From the perspective of
neuroscience, however, there is a much more powerful lesson:
We're all like Alice all the time—our brains must process strange
new information arising from unpredictable experiences every
single day, and provide us with useful responses—except that we
didn't have to drop through the rabbit hole. We're already deep
inside it.

My goal in *Deviate* is to reveal the hidden wonderland of your
own perception to you as my more than 25 years of research
have revealed it to me. You don't have to be a so-called "science
person." Although I'm a neuroscientist, I'm not just interested
in the brain only, since neuroscience is so much bigger than just
the brain. When neuroscience is applied outside the disciplines it
is traditionally associated with—such as chemistry, physiology,
and medicine—the possibilities are not just immense, but

fantastically unpredictable. Neuroscience—when defined more broadly—has the potential to impact everything from apps to art, web design to fashion design, education to communication, and perhaps most fundamentally, your personal life. You're the only one seeing what you see, so perception is ultimately personal. Understanding of the brain (and its relationship to the world around you) can affect *anything*, and lead to startling deviations.

Once you begin to see perceptual neuroscience this way, as I did several years ago, it becomes hard to stay in the lab . . . or at least the more conventional, staid conception of what a "lab" is. So, a decade ago I began redirecting my energies toward creating brain-changing, science-based experiences for the public: experiment as experience . . . even theater. The theme of one of my first installations at a leading science museum was *Alice in Wonderland*. The exhibit, much like Lewis Carroll's strange, topsy-turvy novel, took visitors through illusions intended to challenge and enrich their view of human perception. This first exhibit—which I created with the scientist Richard Gregory, a hero in perception who shaped much of what I (and we) think about the perceiving brain—grew into many other settings, all of them based on the belief that to create spaces for understanding we need to consider not only how we see, but *why* we see what we do. To this end, I founded the Lab of Misfits, a public place open to anyone where I could conduct science "in its natural habitat," a playful and rule-breaking ecology of creativity. This was most dramatically the case when we took up residency in the Science Museum in London.

My Lab of Misfits has enabled me to bring together primatologists, dancers, choreographers, musicians, composers, children, teachers, mathematicians, computer scientists, investors, behavioral scientists, and of course neuroscientists in a place where concepts and principles unite, where the emphasis is on innovation, and where we passionately investigate things we care about. We've had an official "Keeper of the Crayons" and "Head Player" (not *that* kind of player—as far as we know). We've published papers on nonlinear computation and dance, bee behavior and architecture, visual music, and the evolution of plant development. We've created the world's first Immersive Messaging app that enables gifting in physical space using augmented reality, which allow people to re-engage with the world. We've initiated a new way to interact with the public called NeuroDesign, which combines those who are brilliant at telling stories with those who understand the nature of the stories the brain desires. We have created an education platform that, with the raison d'être of encouraging courage, compassion, and creativity, doesn't teach children about science but makes them scientists, and has resulted in the youngest published scientists in the world (and the youngest main-stage TED speaker). Many of the ideas in *Deviate* were created, prototyped, and embodied through experience in this physical and conceptual "Lab of Misfits" space. This means the book is also a product of all these misfits, the interactions between them, and even more significantly, our interactions with historic and contemporary misfits outside the lab.

This brings me to a key theme in the pages ahead: that perception isn't an isolated operation in our brains, but part of an ongoing process inside an *ecology*, by which I mean the relation of things to the things around them, and how they influence each other. Understanding a whirlpool isn't about understanding water molecules; it's about understanding the interaction of those molecules. Understanding what it is to be human is about understanding the interactions between our brain and body, and between other brains and bodies, as well as with the world at large. Hence life is an *ecology*, not an environment. Life—and what we perceive—lives in what I call "the space between." My lab, and all my research on perception, draws on this inherent interconnectedness, which is where biology, and indeed life it-self, lives.

Now I have started all over again and built my lab into a book—hopefully a delightfully misfit one, shot through with deviations. This creates a sense of danger, not just for me but for you as well, since together we will need to question basic assumptions, such as whether or not we see reality. Stepping into such uncertainty isn't easy or simple. On the contrary, all brains are deathly afraid of uncertainty—and for good reason. To change a historical reflex will have unknown consequences. "Not knowing" is an evolutionarily bad idea. If our ancestors paused because they weren't sure whether the dark shape in front of them was a shadow or a predator, well, it was already too late. We evolved to predict. Why are all horror films shot in the dark? Think of the feeling you often have when walking through a familiar forest at night as compared to during the day. At night you can't see what's around you. You're uncertain. It's frightening, much like the con-stant "firsts" life presents us with—the first day of school, first dates, the first time giving a speech. We don't know what's going to happen, so these situa-tions cause our bodies and our minds to react.

Uncertainty is the problem that our brains evolved to solve.

Uncertainty is *the* problem that our brains evolved to solve.

Resolving uncertainty is a unifying principle across biology, and thus is the inherent task of evolution, devel-opment, and learning. This is a very good thing. As you will have observed from experience, life is inherently uncertain because the world and the things that constitute it are always changing. And the question of uncertainty will become an increasingly pressing issue in all parts of our lives. This is because, as we and our institutions become more interconnected, we become more interdependent. When more and more of us are connected to each other, the effects of the metaphorical butterfly flapping its wings on the other side of the world are more quickly and more powerfully felt everywhere, increasing the pace of change (which is at the heart of a nonlinear, complex system). An increasingly connected world is also inherently more unpredictable. This creates fundamental challenges for living today, from love to leadership. Many of the most sought-after jobs today, from social media expert to web designer, weren't even around twenty years ago. A successful company, a thriving relationship, an environment free of dangers—the existence of these things today doesn't guarantee their continued existence tomorrow. You are never truly "untouched" in a connected, flux-filled world. There will always be events that blindside you, that you didn't predict, from the unforeseen change in weather spoiling your BBQ in London on a Saturday afternoon to those in London suddenly finding themselves living outside the European Union. This is why our brain evolved to take what is inherently uncertain and make it certain . . . every second of every day. The biological motivation of many of our social and cultural habits and reflexes, including religion and politics, and even hate and racism, is to diminish uncertainty through imposed rules and rigid environments . . . or in one's vain attempt to disconnect from a world that lives only because it *is* connected and in movement. In doing so, these inherited reflexes—by design—prevent us from living more creative, compassionate, collaborative, and courageous lives. With the making of this kind of certainty, we lose . . . freedom.

At Burning Man in 2014, I had an experience that has stayed with me—actually quite a few, but I'll share this one here. It was a profound—and profoundly simple—example of how deviating can radically change one's brain. As many know, Burning Man is a weeklong festival every August in the Nevada desert that brings together art, music, dance, theater, architecture, technology, conversation, and nearly 70,000 human beings. Costumes are ubiquitous—and at times a complete lack thereof (though often with body paint). It is a city-sized circus of free-form creativity . . . picture a giant pirate ship sailing along on wheels . . . that explodes on the desert floor, then vanishes seven days later, leaving absolutely no trace . . . an essential part of the Burning Man ethos.

On a windy day midway through the week, my partner Isabel and I were riding our bikes and getting to know the "city." Desert dust swirled, silting us and our goggles in a fine layer of beige. We ended up in a camp of people from a town on the southern edge of the Midwest and met a guy I'll call Dave. This was Dave's first year at Burning Man, and he said it was turning out to be a transformative experience for him. At first I internally rolled my eyes at this. Being "transformed" at Burning Man has become not just a cliché but almost an imposed aspiration. If you don't *transform* there, then you have somehow failed. But what is transformation? Of course, no one really knows because it is different for every person, which is why so many people at Burning Man hungrily chase signs of it all week, going around asking: "Have you been transformed?"

The more we talked to Dave, though, the more I realized he really was undergoing a deep shift in his perceptions of self and other. He was a computer programmer from a place with fundamentalist religious values and a narrow outlook on what was socially acceptable. In his town, you either learned to fit in or you were ostracized. Dave had learned to fit in . . . the business casual attire he wore at Burning Man reflected this. But it had clearly curtailed the possibilities of his life, curiosity, and imagination.

Yet here he was, at Burning Man! It was the decision to *be there* that mattered. It was his choice . . . his intention enacted . . . to come, and the questioning manner he had brought with him.

As we stood there in his camp, he told us that the little green plastic flower that I saw stuck behind his ear—perhaps the least flamboyant adornment in Burning Man history—had provoked an epic struggle inside him. He had sat in his tent for two hours that morning weighing whether or not to wear the flower. It had forced him to confront a complex host of assumptions in his mind—about free expression, masculinity, aesthetic beauty, and social control. In the end, he gave himself permission to question these assumptions symbolically manifested in a plastic flower, and stepped out of his tent. He seemed both pleased and uncomfortable, and in my eyes far more courageous than most of the people out there in the Nevada desert that day in search of something powerful.

As a neuroscientist, I knew that his brain had changed. Ideas and actions previously out of his reach would now be available to him if he was willing to question his assumptions, and in doing so create a new, unknown terrain of wondering. As a person, I was moved.

This is what transformation looks like: Deviation *toward* oneself. So simple. So complex.

Nothing interesting ever happens without *active* doubt. Yet doubt is often disparaged in our culture because it is associated with indecision, a lack of confidence, and therefore weakness. Here I will argue exactly the opposite. That in many contexts, to "doubt yet do . . . with humility," like Dave, is possibly the strongest thing one can do. Doubt with courage and your brain will reward you for it through the new perceptions this process opens up. To question one's assumptions, especially those that define ourselves, requires knowing that you don't see the reality—only

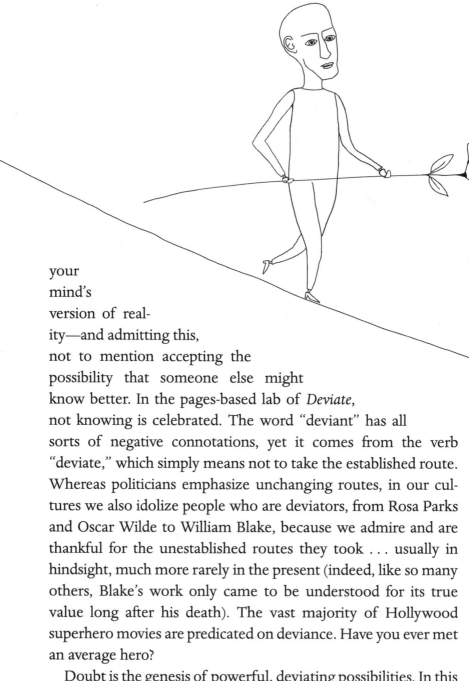

your
mind's
version of real-
ity—and admitting this,
not to mention accepting the
possibility that someone else might
know better. In the pages-based lab of *Deviate*,
not knowing is celebrated. The word "deviant" has all
sorts of negative connotations, yet it comes from the verb
"deviate," which simply means not to take the established route.
Whereas politicians emphasize unchanging routes, in our cultures we also idolize people who are deviators, from Rosa Parks
and Oscar Wilde to William Blake, because we admire and are
thankful for the unestablished routes they took . . . usually in
hindsight, much more rarely in the present (indeed, like so many
others, Blake's work only came to be understood for its true
value long after his death). The vast majority of Hollywood
superhero movies are predicated on deviance. Have you ever met
an average hero?

Doubt is the genesis of powerful, deviating possibilities. In this
way, the human brain is able to shed constricting assumptions
and see beyond the utility with which the past has trained it to
see. As I like to say, *the cash is in the questions.*

Be Delusional

The doubt-driven ride this book will take you on is going to *physically change your brain*. This isn't braggadocio, but a fact-based understanding of everything from the electric patterns of your thoughts to the neurons of your emotions. The simple act of reading can change your brain because two and a half decades of research have led me to one indisputable conclusion: what makes the human brain beautiful is that it is *delusional*.

I'm not talking about insanity. What I'm getting at has to do with the brain's imaginative powers of possibility and how richly they interact with behavior. We can all hold *mutually exclusive* realities in our minds at the same time, and "live" them out imaginatively.

Human perception is so layered and complex that our brains are constantly responding to stimuli that aren't real in any physical, concrete sense, but are just as vitally important: our thoughts. We are beautifully delusional because *internal* context is as determinative as our external one. This is verifiable at the neural level: fMRIs (functional magnetic resonance imaging, a technique for tracking brain activity through blood flow) show that an imagined scenario lights up brain regions the same way the real-life equivalent scenario does. In other words, ideas and thoughts and concepts have lives inside of us. They are our history, too, and directly feed our current and (maybe more importantly) future behavior. As such, our perception is much more plastic and subject to influence than we're often aware of or comfortable admitting. The stock market tends to go up when it's sunny, and down when it's not. The seemingly rational decisions we make, then, are actually guided by "invisible" forces of perception that we're not even conscious of.

Another example: In 2014 the Lab of Misfits organized our first party/study, an initiative we call The Experiment, which is designed to do many things. One is to improve the quality of scientific research by taking it out of the artificial situation of a lab and into authentic human situations. The situation we engineered was a true social gathering in which people ate and drank and talked in an old crypt with strangers with a larger theatrical context. For the participants it was designed to be purposefully ambiguous as to whether it was science, a nightclub, an interactive theater and/or cabaret, but it was a memorable experience in which they also served as subjects in an experiment-as-experience. The goal of The Experiment is to discover, challenge and raise awareness through "empirical embodiment" of what it is to be human. One of our experiences sought specifically to see whether people group themselves depending on how they perceive themselves as either powerful or not.

After the food, once everyone was full, relaxed, and enjoying themselves, we had people do a brief writing exercise to prime them into a perceptual state. Depending on the memory they were prompted to recall, they were primed into either a low-power state, a higher-power state, or a neutral-power state. What this means is that their recollection prompted them to unconsciously perceive themselves to be either less or more in control. We then had them walk in a large concentric circle within a big underground crypt space in a Victorian jail in East London. Next, we asked them to segregate themselves under two lights at opposite ends of the room—in short, to stand next to the people who "feel like you." That's all we said.

What happened shocked the guests as much as it did us scientists. Without knowing who had been primed in which way, *the people organized themselves according to their power-state with two-thirds accuracy.* This means that well over half the people in each corner were with other people "like themselves." This was astounding for two reasons: One, it showed how strongly the participants' simple thoughts about themselves changed their own behavior; that is, their imagining changed their perceptual responses. Two, the people somehow perceived the imaginatively primed perceptions of others. What a wondrous example of how delusions affect not only our behavior, but the ecology in which we interact as well. In the chapters ahead, you will learn how to make your brain's delusional nature enhance your perception.

I want to create a fresh layer of meaning in your brain that will be as real as anything else that has affected your perception—and your life. The narrative of this book embodies the process I'm going to teach you. I constructed it so that reading from first page to last *is* seeing differently. It will allow you to experience what creativity feels and looks like from the inside. Think of it as a software solution for your perception. When you're done with the book, you simply change the context and reapply the software. Perhaps the most encouraging part is that you don't have to acquire a new base of knowledge.

To fly a plane, you first have to train as a pilot, which involves a tremendous amount of specialization and practice. But in order to deviate into new perceptions, you already have the basics. You don't have to learn to see and perceive. It's an essential part of who you are, if not *the* essential part. In this sense, you already have a firsthand account of the subject of this book. Furthermore, the process of perception is the same process by which you change perception. This means you are your own pilot (in the context of your larger ecology). My task is to use the science of your brain to teach you a new way to fly, and to see anew what you thought you had already seen.

One of the ways I will do this is by applying my knowledge of perception *to your reading experience*. For example, the brain thrives on difference . . . on contrast, since only by comparing things is it able to build relationships, which is a key step in creating perceptions. This is why you will find deviant design elements, such as varying font sizes and occasionally puzzling images. On the page you will also find exercises, tests, and self-experiments that require your participation. (They won't be tedious; in one I'm going to make you hold your eye open and momentarily "go blind.") When I began *Deviate*, I wanted to challenge assumptions about what a science book could be—my own step into uncertainty. What better space to do this than in a work about innovation and the brain, using the brain as my guide? This book is different in other ways as well.

In my view, as soon as you've told something to someone, you've taken the potential for a deeper meaning away from them. True knowledge is when information becomes embodied understanding: We have to act in the world to understand it. This is why *Deviate* will not give you recipes. Instead of a how-to guide that offers task-specific formulas, I will give you principles that transcend any single context. Just because you are able to make one fantastic meal by following a recipe doesn't mean you are now a great cook; it means you are good at following the instructions of a great cook. While it may have worked once, it hasn't given you the wisdom to make *your own* fantastic meal, as you have no idea why the recipe is a good one. *Understanding* why the recipe is a good one (and how) is one key aspect of what makes a chef a chef.

Deviate is designed to innovate your thinking by giving you new awareness, which creates the freedom to change. The first half will explore the mechanics of perception itself, making you reconsider the "reality" you see and helping you to know less than you think you know now. Yes: that is my aim, for you to actually know *less* overall, while understanding more. The second half will then make this understanding practical by giving you a process and technique to deviate in your life.

When you finish this book, I have only one true hope: that you will embrace the perceptual power of doubt. This book is about celebrating the courage of doubt, and the humility that comes with understanding your own brain. It's about why we see what we do, and how recognizing that we don't have access to reality leads us to get more things right. Which is all just another way of explaining why I wrote this book: so that you too will be a misfit.

CHAPTER 1

Being in Color

When you woke up this morning and opened your eyes for the first time, did you see the world accurately, the way it really is? If you said no, let me ask the question a different way: Do you believe in illusions? Most of us do. If so, then by definition you believe that the brain evolved to see the world accurately, at least most of the time, given that the definition of *illusion* is an impression of the world that is different from the way it really is. And yet we don't see the world accurately. Why? What is going on inside our complex brains (or more precisely, in the complex interaction between our brain and its world) that makes this so? First, however, we must address an urgent empirical question and satisfy that human need to "see it with my own eyes": Where is the proof that we don't see reality? How can we *see* that we don't see it? The answer to this question is where we begin to dismantle our assumptions about perception.

In February of 2014, a photo posted on Tumblr went viral on a global scale and inadvertently spoke to just this issue of the subjectivity of perception—and spoke loudly. The questions it raised about what we see generated thousands more questions across Twitter and other social media and on TV, as well as in the minds of people who kept their astonishment private. You may or may not have encountered the photo, but if you did you'll remember the image itself gave the phenomenon its name—*The Dress*.

It all started with a wedding in Scotland. The mother of the bride had sent a photo of the dress she was going to wear to her daughter: a straightforward gown of blue fabric with stripes of black lace running across it. Yet the photo itself was anything but straightforward for perception. The bride and groom couldn't agree on whether it was white with gold stripes or blue with black stripes. Baffled by their disagreement, they forwarded the image to people they knew, including their friend Caitlin McNeill, a musician who was performing at the wedding. She nearly missed her stage call because she and her bandmates (who, like the couple, didn't see the dress the same) were arguing about the image.[1] After the wedding, McNeill posted the slightly washed-out photo on her Tumblr page with this caption: "guys please help me—is this dress white and gold, or blue and black? Me and my friends can't agree and we are freaking the fuck out." Not long after she published this short commentary, the post hit viral critical mass and, as the saying goes, the photo "broke the Internet."

Over the following week The Dress ran its course as most viral phenomenon do, with the explosive, out-of-nowhere virality becoming as much the story as the item—in this case a simple photo of a piece of clothing—that instigated it. Celebrities tweeted and feuded about it, reddit threads proliferated, and news organizations covered it. Those of us who research color were suddenly inundated with requests for interviews, as it seemed everyone wanted to know why they saw the colors differently. Even the usually sober *Washington Post* published the sensationalist headline: "The Inside Story of the 'White Dress, Blue Dress' Drama That Divided a Planet."[2] Yet in spite of the overheated excitement and debate, people were having an important conversation about science—to be precise, *perceptual neuroscience.*

I found this remarkable on several levels, but most profoundly in that it hinted at the way meaning is a plastic entity, much like the physical network of the brain, which we shape and reshape through perceptual experiences. Understanding this, as we'll see in later chapters, is the key to "re-engineering" your perceptual past to liberate unforeseen thoughts and ideas from your brain cells. The Dress phenomenon was a perfect example of how meaning creates meaning (as news agencies around the world started reporting a story largely because it was being *reported* elsewhere and therefore assumed to be meaningful, and thereby making it meaningful), which is a fundamental attribute of perception itself. But I was also struck by the fact that it wasn't the illusion per se that was grabbing people's attention, since we are accustomed to them (though usually as simple "tricks"). What seemed to grab people was that they were seeing it differently from each other. We are very familiar with having different conceptual views about things, though. So how and why was this different? It came down to this: it was because it was color.

We are OK with speaking a different language than another person, but when my friends, loved ones, and others whose perception and grasp of reality I trust differ at such a fundamental level as color, this raises . . . for a beautiful moment, albeit too brief . . . deep, largely unconscious existential questions about how I see the world around me. It unnerved something at the core of how people understood their very consciousness, selves, and existence. As the actor and writer Mindy Kaling tweeted on February 25th, in the midst of the frenzy of #TheDress (one of her many impassioned tweets about it): "I think I'm getting so mad about the dress because it's an assault on what I believe is objective truth."

This is the crux about perception and self that The Dress brought up for so many: there is an objective "truth" or reality, *but our brains don't give us access to it.* We got a shocking "look" at the cracks in our highly subjective reality through the photo of the dress—and it was a bit upsetting, or at least unsettling. The key for understanding how to enhance creativity through an understanding of perception . . . as you will soon see . . . is the following: this brief step into the most fundamental of uncertainties was also *exciting* for people. It freaked them out a bit yes, but it thrilled them, too.

For me personally, it was even more exciting, since I got to observe in real-time as millions of people took a dramatic step forward in understanding. Yet some people also wrote off The Dress: *"OK, my perception didn't see reality this time, but it usually does."* "No!" I wanted to shout. "You never, ever see reality!" Unfortunately, this essential point never became the central "story" of The Dress, though some in the scientific community took advantage of the opportunity to engage a broader audience with a topic that in any other cultural moment would have seemed abstruse and irrelevant. In May 2015, for example, *Current Biology* simultaneously published three studies of The Dress. One found that the distribution of colors of the dress corresponded to "natural daylights," which makes it harder for your brain to differentiate sources of light from surfaces that reflect light (more on this in the next chapter). Another study made a discovery about how the brain processes the color blue, revealing that things have a greater probability of appearing white or gray to the human eye when "varied along bluish directions." The last study, which surveyed 1,401 participants, found that 57 percent see the dress as blue/black, while the white/gold perception of the dress was more common in older people and women. Additionally, on a second viewing, participants' perception sometimes switched white/gold to blue/back, or vice versa. In short, the viral photo proved to be an ideal experimental object to further the study of visual perception.[3]

Still, none of this answers the question of why the hell people saw that dress differently.

#TheDress tapped into not only how perception works, but why it matters so much to us. It illustrates the extremely counterintuitive nature of our brain. If we saw the world as it really is, then things that are the same should look the same. Likewise, things that are different should look different . . . always, to everyone. This seems sensible and correct, something we can count on from our perception (or so we thought). After all, seeing different intensities of light is the simplest task the visual brain performs, so simple that even some jellyfish can do it—and they don't even have a brain.

Yet perceiving light is not actually as straightforward as it may seem, despite the fact that we do it every waking millisecond. The billions of cells and their interconnections devoted to it are evidence of its difficulty. We rely on this perceptual skill to make instinctive decisions that serve us as we move through the world. The Dress, however, revealed that just because we *sense* light, we don't necessarily *see* the reality of it.

In the first of the images that follow, each circle is a different shade of gray. It's easy to perceive the varying gradations. Things that are different should look different, and they do.

In the second image we are looking at two circles of identical shades of gray.

Now look at the third image. The gray circle on the left inside the dark box looks lighter than the gray circle on the right inside the white box. They appear to be two distinct shades of gray.

But they're not. They are the exact same gray.

This is the objective reality—which is fundamentally different from our perceptional reality. What's more, every reader of this book will perceive these three images the same way you did, with the same departure from the physical reality of what is printed on the page. What is more, it is not simply that dark surrounds make things look lighter than light surrounds. The opposite can also be true: light surrounds can make things look lighter and dark surrounds can make things look darker as shown in the fourth image, where the central regions, which look like circles obscured by four squares, appear differently light in the way just described.

But the most powerful message here is one that never came up during the viral Dress phenomenon. What is true for vision is indeed true for *every one of our senses*. What people realized about the subjectivity of their sight also goes for every other facet of their "reality": there are illusions in sound, touch, taste, and smell, too.

A very well-known example of tactile "gaps" between perception and reality is called the "Rubber-Hand Illusion." In this so-called trick, a person is seated at a table with one hand resting in front of them, while the other is out of sight behind a divider. A fake hand is set down in front of the person in place of the out-of-sight hand, so they have what looks more or less like their two hands resting on the table, except that one of them isn't theirs (which of course they're aware of). Then the "experimenter" begins lightly brushing the fingers of the hidden real hand and the fake visible hand at the same time. Sure enough, the person immediately starts identifying with the fake hand as if it were theirs, feeling as if the brushing sensation is happening not behind the hidden divider, but on the fake hand they suddenly feel connected to. For the purpose of perception, that hand becomes real!

The Rubber-Hand Illusion is what's called a *body transfer*, but our brain's way of processing reality . . . rather than giving it to us directly . . . also opens us up to other slightly freaky "mix-ups" of the senses. Researchers have shown, for example, that we are capable of hearing "phantom words." When listening to nonsense sounds, our brains pick out clear-seeming words that aren't actually there in the audio. There is also the "Barber's Shop Illusion," in which a recording of snipping scissors gives the impression of the sound getting closer or farther depending on the volume rising or lowering, when the sound isn't changing position at all. Or think of the common experience of sitting in a stationary car or plane, and when the car or plane next to you

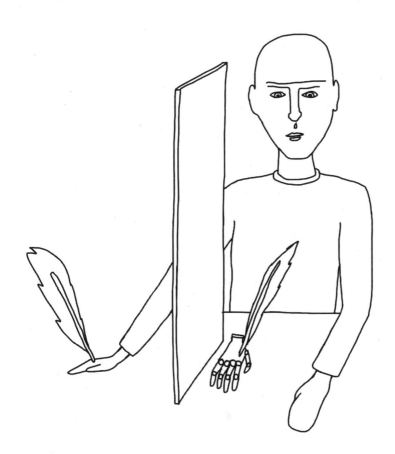

begins to move, at first you think you are the one in motion. There are many, many more phenomena like these.

One of the first people to pick up on what seemed a "kink" in visual perception was Johann Wolfgang von Goethe, the eighteenth-century man of letters we know today as the father of modern German literature. In his day he was famous (though also slightly infamous, as you'll see) as an omnivorous dabbler across disciplines, as likely to throw his energies into osteology (the study of bones) as he was to immerse himself in botany. While literature was his first love, Goethe was above all else a man of passion, often described as being wild in nature, so much so that as a young man his friends referred to him as the "wolf" and the "bear."[4] (In his college days in Leipzig, his out-of-fashion Frankfurt dress style also amused his peers.) But he was able to tame this wolfish unruliness into a high-society charisma as he became a literary celebrity in his twenties, the Duke Karl August soon appointing him to several state roles, including Director of the War Department. Goethe's headstrong and even reckless hunger for new intellectual experiences returned in full force in the late 1780s when his *"manysidedness,"* as a biographer once labeled it, led him to study light and color.

Goethe had recently spent a very happy two years in Italy. There he had become acquainted with the German painter Johann Heinrich Wilhelm Tischbein while also exploring his own talents in the fine arts. He eventually accepted that he had none, but he returned to Germany with a reignited interest in the natural world artists sought to capture. "No one acquainted with the charm which the secrets of Nature have for man, will wonder that I have quitted the circle of observations in which I have hitherto been confined," he wrote in an unpublished essay. "I stand in no fear of the reproach that it must be a spirit of contradiction which has drawn me from the contemplation and portraiture of the human heart to that of Nature. For it will be allowed that all things are intimately connected, and that the inquiring mind is unwilling to be excluded from anything attainable."

This declaration led to one of history's most legendary examples of a literary personage disastrously elbowing his way into the halls of science. Depending on who you ask, what transpired was either a rewardingly poetic misadventure into a realm the belletrist innocently misconstrued, or an arrogant and willfully wrongheaded foray into a field where he did not belong. In reality it was a mixture of the two—neither and both. Goethe's passionate way of being wasn't well suited to the cold-hard-facts approach of science, yet his writerly eye for the revelatory moment did play a key role in his scientific forays when he had his own moment, albeit a *falsely* revelatory one.[5]

Drawn to optics, Goethe had borrowed a prism to test Newton's groundbreaking work that refracted white light into its constituent colors, a discovery that today would have likely won Newton a Nobel Prize.[6] Executing the experiment incorrectly, and far from fully versed on the theory behind it, Goethe expected to beam the full spectrum of color on the wall of his home, only to beam nothing. The wall remained a white blank on which he could project his rapidly growing conviction: "Newton's theory is false!"

Ardently sure that contemporary science was terribly off course in understanding light, he took leave of his diplomatic responsibilities to devote himself to physics. Scientists of the time ridiculed him while littérateurs and nobles cheered him on, certain that the wild wolf of a poet would dethrone Newton. The Duke of Gotha gave him a laboratory; a prince sent him new, better prisms from abroad. Then in 1792, while conducting his inquiry, Goethe noticed that a white light could produce shadows *with colored hues*. Likewise, colored lights changed hues depending on the *"opaque,"* or semi-transparent, medium they passed through. A yellow light could redden all the way to ruby through the opaque. This too appeared to violate Newton's physical laws purporting to explain light, and the disconnect created a still deeper chasm in Goethe's understanding of Newton's theories that purported to explain reality. So he narrowed his focus to work out this problem and was sucked into a twenty-year obsession with color and perception.

In works like *The Sorrows of Young Werther*, the tumultuous story of a young man's unrequited love, Goethe proved himself to be a poet of the human interior, so perhaps it's no surprise that when it came to color he at first wasn't always able to step outside the cave of his own perception to see that the disjuncture was inside. Like most of us, he took for granted that he saw reality; after all, his prodigious mind had allowed him to "see" human realities so clearly in his writing, not to mention that perception was well over a century away from growing from a concept into a locus of science. Soon enough, however, he gave up the idea that the seemingly mismatched hues were due to some physical quality of light that science had yet to explain. Instead, he gradually realized that the colored appearance of certain shadows was a result of human perception's interaction with its environs—not a mystery of the world but a mystery of the mind.[7] Yet he could only uselessly paw at the reasons *behind* this strange mystery inside his own brain, so he became a scrupulous recorder of every phenomenon of light he observed.

The subject of color consumed Goethe to the point that, in 1810, he published a massive tome on his investigations called *Zur Farbenlehre*, or *Theory of Color*. The "science" of the book has long since been dismissed, most notably his attacks on Newton. Goethe's taxonomic opus did, on the other hand, incite intense philosophical discussion, leading Wittgenstein to pen *Remarks on Color* and Schopenhauer to write *On Vision and Colors*. Yet Goethe's encyclopedic descriptions of color, like all his work, still make for lyrical, pleasing reading today: "These colors pass like a breath over the plate of steel; each seems to fly before the other, but, in reality, each successive hue is constantly developed from the preceding one."[8] His sense of wonder is an invaluable trait we can still learn from. And this obsession in Goethe's life may explain the origins of the famous line from his masterpiece *Faust*, when Mephistopheles, a stand-in for the devil, says to the easily corruptible Student: "Grey is, young friend, all theory."[9]

It is worth taking a moment to dwell on the word *reality* as Goethe used it at the time of writing *Theory of Color*, as the eighteenth century became the nineteenth. This was still the heyday of the Enlightenment, a revolutionary period in which Western societies were trading in the superstitions of the medieval age for a newly cemented belief in the rationality of man. If "man" was a being of capital-R Reason, then wouldn't it have been a contradiction to suggest that the very sine qua non of reason—perception—prevented humans from seeing reality accurately? Humans were rapidly conquering their time and place with political theory, criminal law, and mathematical proofs, so how could the minds producing such important changes in reality not see reality itself? At its core the story of Goethe's glorious boondoggle with color says less about him (though it certainly does say a lot about him) than it does about the assumptions that reigned at a time when so much about the brain was still unknown—the very same assumptions about "seeing reality" that inflamed so many people online two centuries later with The Dress. Because when it comes to understanding perception, the correct assumption is deeply counterintuitive.

Most people assume we see the world accurately as it really is, even many neuroscientists, psychologists, and cognitive scientists, because, well, why wouldn't we? At first "glance," it seems a wholly bad idea to see it in any other way. The seemingly logical premise that we see reality, however, doesn't take into account a fundamental fact about ecology and how our minds actually operate therein, thus missing the essential truth that *our brains didn't evolve that way*. But then what *did* our brains evolve to do? They evolved to survive—and that's it.

A key nuance to understand is this: Despite the fact that your brain doesn't perceive reality, your senses aren't in any way "weak." The human brain was shaped by the most rigorous, exhaustive research and development and product-testing process on our planet—*evolution* (and subsequently development and learning). As such, the test with the gray circles wasn't meant to demonstrate how easy it is to play tricks on the senses. Quite the opposite (though we'll wait until Chapter 4 to learn why the illusion you saw wasn't actually an illusion). For now, just rest assured that evolution (and development and learning) doesn't produce fragile systems, which is why changing the way you perceive is possible—because "fragile" isn't at all the same thing as "malleable" and "adaptable." Evolution's "aim" is resilience, robustness and to be, well, evolvable. The human species is a superb example of this process at work. Which means when you look out into the world, you're actually looking *through* millions of years of history.

So you didn't evolve to see reality . . . you evolved to survive. And seeing reality accurately isn't a prerequisite to survival. Indeed, it could even be a barrier to it. Without this as your founding premise about perception, you will be stuck in old ways of seeing, since if you attack a problem with the wrong assumption, there is nowhere to go but deeper into that

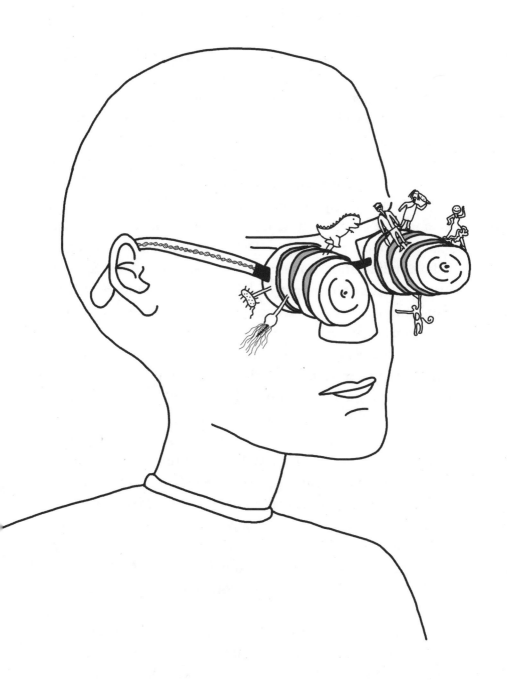

assumption, whether you know you're getting further from the truth or not.

There is so much in the stories of Goethe and The Dress that is at the heart of seeing differently: the ability *to challenge the prevailing assumption* (accidentally or otherwise) that constrains the brain's search for the answer to the wrong place. Goethe became so frustrated, much like Mindy Kaling, because he took it for granted that his perception gave him direct access to reality, not that his brain was simply a very sophisticated interpreter. The wonderfully epigrammatic Goethe nevertheless produced a memorable phrase that he likely used to console himself when his perception failed him: "The man with insight enough to admit his limitations comes nearest to perfection." Perhaps this is true, but even the insight of knowing and admitting that you don't know can be a challenge. This reminds me of a well-known joke in science:

Imagine a very dark street. In the distance is a single streetlight illuminating a small circle of the sidewalk (all other lights along the road are for some reason off). Within this circle of light is someone on their hands and knees. You walk up to them and ask them what they're doing. They reply: "Looking for my keys." Naturally you want to help as they seem really quite desperate. It's late and cold and surely two people looking in the same space is better than one. So for reasons of efficiency—to help you guide your search—you ask: "By the way, where did you drop them?"

The answer: "I dropped them way over there," says the person, pointing to a place in the dark a hundred meters behind you.

"Then why in the hell are you looking here?" you ask.

"Because this is the only place I can see!"

Our assumptions create the light in which we are able to see differently—or not, depending on how strongly we avoid exploring the shadows where keys to new paths may be hiding.

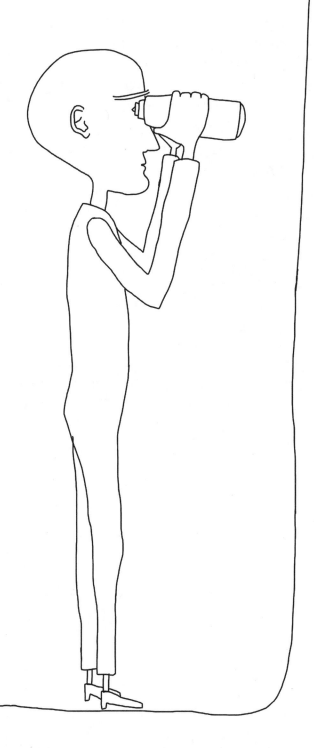

Which is why I love this story so much. It's a parable of sorts, a story-as-warning that forces you to reflect about your ways of looking. And it perfectly sets up the starting point for this book. If we try to "decode" how perception works based on the assumption that we evolved to see reality accurately, we will not only fail to explain the brain, but we'll never be able to have a more direct approach to seeing differently. If, however, we start with a different assumption . . . even one that may challenge deeply personal assumptions about the world and ourselves . . . then this book will "translate" a new way of thinking, seeing, and perceiving. But it's essential that our new assumptions are founded on the startling discoveries of neuroscience and not preconceptions about what feels real and what doesn't. Only by accepting these strange truths can we open a new world of perception.

But a question remains: If our brains are so highly evolved, then *why* don't we have access to reality? In the next chapter I'll answer this question by showing you that all information means absolutely nothing in and of itself, including these very words you're reading. From there, we will explore the brain mechanics behind *not* deviating and seeing reality . . . because only then can you begin to perceive what becomes possible.

CHAPTER 2

Information Is Meaningless

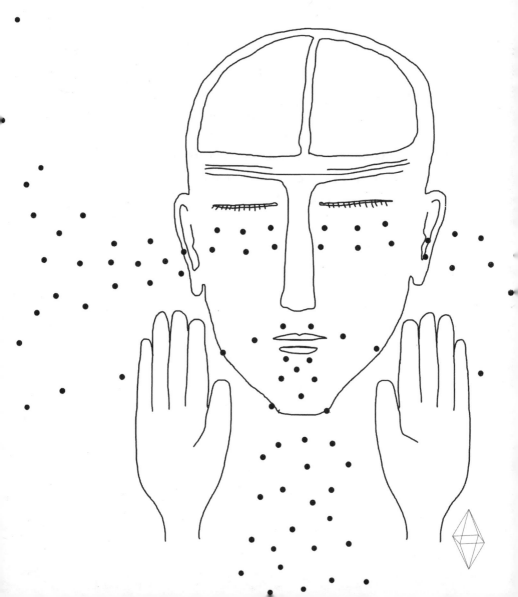

We live in the Wiki Age, an era in which information moves and grows more freely than anyone could have imagined even in the futuristic-sounding year of 2001, back when the Internet was beginning to reshape our lives. When we argue over a point with friends at a meal we pull out our phones and resolve the question in seconds. We don't get lost because it's very hard to be lost these days with GPS. Our social networks span far beyond the people we truly know (or even care to know). Information is overflowing and we're hungrily drinking in each terabyte ... mapping every street, archiving every tweet, capturing every passing moment with our phones. It is the Age of Reason pushed into the new frontier of the Digital Age. This unprecedentedly accessible, ever-expanding web of data has altered our daily existence, and yet very little of it is translating into *new understanding*. This is because when it comes to creativity, success, and even personal happiness, information *isn't* itself power. This is true even—or especially—at the level of our senses.

To
understand
human
perception,

you
must
first
understand
that
all
information
in
and
of
itself
is

meaningless.

The reason for this is simple: The information from the world that falls onto our different senses could literally mean anything. It is nothing more than energy or molecules. The photons entering our eyes, the vibrations through the air that enter our ears, the breaking of the bonds of molecules that creates friction across our skin, the chemicals that land on our tongues, and the compounds that enter our noses—all are just electro/chemical energy of one kind or another. These are the elements that emanate from our physical world—real reality, as it were. Yet we don't have direct access to those sources of energy, only to the waves of energy and gradients of chemicals that they produce. We sense the *changes* in stuff, not the stuff itself. It would be useless to have direct access to the "stuff" because in isolation it would mean absolutely nothing . . . much in the same way that a single water molecule doesn't tell us about whirlpools. Information doesn't come with instructions.

Instead, the "reality" that our perceptions see is the meaning of the meaningless information that your brain receives . . . the meaning your ecology *gives* it. It is critical to understand that the meaning of the thing is not the same as the thing itself. In other words, perception is similar to reading poetry: you are interpreting what it means, because it could mean anything.

Y U
MA E
TH
EANI G

by interacting with the world (i.e. your ecology). This is true from the color of stoplights to the smiles (or frowns) of strangers you pass on the street. Your brain is such a fast, prodigiously skilled machine of meaning: A certain light means a certain surface color, a certain smell means a certain food, a certain voice means a certain person, a certain touch means a certain emotion, and a certain sight means a certain place. But note that the surface is not in fact colored in any literal sense. The seeing of the color red is itself the seeing of a past narrative of meanings. Such perceptions make it feel like the mosaic of reality lands on your senses with meaning already built in, but in fact no meaning is ever predetermined. Equally, no meaning is meaningless . . . only the raw information is. Now let's learn why information is meaningless, and why our species (and indeed any living system) evolved to have a brain that *creates* a perception of a world, rather than just relays it.

Reality, according to the eighteenth-century Irish philosopher and Anglican bishop George Berkeley, is simply "ideas . . . imprinted on the senses."[10] Was he correct about this?

Born in 1685, Berkeley was a metaphysician and religious man, though from the vantage of history we could think of him as a theoretical neuroscientist before the field ever existed. He was a product of the early Enlightenment, a thinker for whom faith and science weren't in conflict; critical reason was the vehicle of his belief in God and not an impediment to it. Yet because of his divisive views on human perception, with which he was obsessed, Berkeley has never attained the historical stature of, say, a Nietzsche or Hegel. Nevertheless, Berkeley's insights into the human mind are profound, and his life was one of industrious dignity. Alongside his work in philosophy and theology, he concerned himself with social causes, establishing programs to help children and vagrants, combating unemployment, supporting local artisans, and planting myrtle and flax. Berkeley was, as one biographer put it, "a bishop in his shirt sleeves."[11]

Berkeley's lifelong philosophical project was an impassioned defense of subjective idealism or empirical idealism—the belief that things exist only as a function of the mind. Back in Berkeley's day, most conceived of the brain as a standalone, ready-made entity that wasn't shaped and reshaped by its interactions inside and outside itself. Yet his works arguing his position on what the human mind perceives manage to hit a truly intuitive space of being both spiritually centered and instinctively scientific.

A Treatise Concerning the Principles of Human Knowledge is one of Berkeley's best-known works and in it he sets down his beliefs on perception. "It is indeed an opinion STRANGELY prevailing amongst men," he writes, "that houses, mountains, rivers, and in a word all sensible objects, have an existence, natural or real, distinct from their being perceived by the understanding. But, with how great an assurance and acquiescence so ever this principle may be entertained in the world, yet whoever shall find in his heart to call it in question may, if I mistake not, perceive it to involve a manifest contradiction. For, what are the forementioned objects but the things we perceive by sense? and what do we PERCEIVE BESIDES OUR OWN IDEAS OR SENSATIONS?" His all-caps emphasis may read like an email or text from an overenthusiastic friend, but three hundred years later we know that he was right: We don't see reality—we only see what our brain gives through what the "space between" yields.

Berkeley went further than neuroscience does and claimed that in fact nothing could have an "existence independent of and without the mind." As a metaphor for subjective human perception, this is a compelling way of framing the issue, since we don't experience the existence of anything outside ourselves except through our brain's (and body's) meaning of it. But taken at face value, Berkeley's "immaterialism" is wrong, since clearly the world exists whether we are there to perceive it or not. When a tree falls in the woods, it does indeed create energy in the form of vibrations in the air. But without anyone / anything there to hear it, that transition in the energy state of the air makes no "sound," though it does have an objectively physical effect. Yet by recasting Berkeley's ahead-of-his-time insights through the apparatus of modern neuroscience, we can identify the four reasons why we don't have access to reality.

1. We Don't Sense All There Is to Sense

Our perception is like being inside of a *mobile home* (not a terribly glamorous metaphor, admittedly, but useful for our purposes). Our senses are the windows of our home. There are five: sight, smell, hearing, touch, and taste. From each window, we obtain different kinds of information (i.e., energies) from the world. Critically, we can never step outside the trailer, though we can move it. But even when we move it around, we're still constrained by the windows. So, clearly, we are limited in our ability to sense the world around us, and you may be surprised to discover that these windows of sensing are actually smaller than you think.

Let's look at light. Light is the narrow range of electromagnetic radiation that is visible to the human eye; it is only a fraction of the electromagnetic spectrum. Light has many properties and one is its *range*, or quantity. The light we see travels at the wavelengths (frequencies) that our retinas and visual cortex are sensitive to. We don't see ultraviolet (UV) light, and we don't see infrared either. We are sensitive to only an incredibly small percentage of the electromagnetic radiation that is out there to sense. New technologies like night-vision goggles are "broadening" our senses, but they don't change our biology. In contrast, the biology of other species embodies much more advanced technologies, enabling them to perceive a far, far more diverse range of light than we can.

Reindeers are best known for their noses, but it is their eyes that are most fascinating. While reindeer aren't able to fly across the night sky pulling Santa Claus's sleigh, they did evolve a superhuman power of sorts (relative to us): they can see ultraviolet light. Why do reindeer have this advantage? It relates to the harsh logic of survival in the inhospitable arctic landscape they inhabit. Being able to sense which surfaces reflect UV rays allows them to perceive which surfaces don't. Lichen, a main source of food for reindeer, doesn't, so what's really at stake here is dinner.[12]

Reindeer's UV vision is essentially a homing device for locating food, similar to a bloodhound's highly developed sense of smell. Beyond reindeer, insects, birds, and fish also visually sense much more than we do. Bumblebees, for instance, have a highly sophisticated color vision system that enables them to sense radiation in the UV. Interestingly, their ability to see color evolved even before the pigmentation of flowers did, which means flowers evolved to look the way they do because they wanted to be "beautiful" (attractive) to bees. Contrary to our highly human-centric view of the world in general, flowers aren't there for our benefit, to inspire, say, the English romantic poets in the hills of the Lake District. Their colors and colorations evolved to be attractive to almost everything *but us*. Birds, too, have twice the number of color receptors in their retinas that we do. In relative terms, we're *vastly* color-deficient compared to them.

Range isn't the only aspect of light we don't see in its full reality. There is also its quality or *orientation*, called polarization. All light is either polarized (the waves of vibrating energy charges occur on a single plane) or unpolarized (the vibrations occur on multiple planes). You and I don't perceive polarity, even if you may own polarized sunglasses that help reduce glare by ensuring that only vertical waves, not reflected horizontal waves, come through. Many animals do, however. Take the stomatopod, or mantis shrimp.

The stomatopod is a quirky, shallows-dwelling crustacean with a body like a lobster tail and eyes swaying on stalks. It has the most complex eyes scientists know of, with what some call "stereo vision," or tuneable sight with eight channels, although even this doesn't capture the staggering power of its vision.[13] Stomatopods have sixteen visual pigments—the substance that turns light into electricity for nerve receptors in our brains— whereas humans have only three. In the killed-or-be-killed world of underwater life in which appearances are fatally deceptive, such a highly developed sense gives the stomatopod an edge when it is hunting (or being hunted). Birds also perceive polarization, which allows them to see the electromagnetic structure of the sky, and not just shades of blue. When they fly through the air, birds (presumably) see tremendous patterns that are constantly changing depending on the angle of the sun relative to their position. These patterns enable them to navigate, since the structure of the sky changes according to the angle of the sun. The birds use this information to decide where to go to next. In other words, to "find their way," they often look up, not down.

So just imagine . . . Looking out onto the world through a bird's perception. When we look up on a clear sunny day we see nothing but a uniform blue. But to birds and bees, the beautiful, cloudless blue sky that we see is never uniform. They see it as an ever-changing complex pattern, a meaningful navigational land-scape composed of shape and structure. What form does this take in their perception? What do they actually see? To try to "envision" it is impossible, as it is a different perceptual reality altogether, like someone without eyes imagining color itself. That is amazing!

To come back to our metaphor of the mobile home, our vision window is tiny, like a miniature porthole, while other species have a wall's worth of glass for their sight, like a floor-to-ceiling window. Of course, this goes for other senses as well. After all, what is a dog whistle but yet another layer of reality we don't have access to?

The differences between us and other animals don't mean we are less evolved for perceiving less. It just means that our evolutions naturally selected differently for survival. We humans have our legendary opposable thumbs, an advantageous evolutionary happenstance that helped us flourish, while stomatopods are happily thumbless, since they evolved in a context with a different set of necessities.

The point is this: We're getting only what our senses give us through the inherently limited windows of our mobile home. While energy is realistic, it is not the reality of objects, distance, etc., even if it is created from fragments of that physical world. But this is only the first of the reasons why information is meaningless.

Pa qui dolupta turehent quos quae ped most, sita ad qui nulliat optatur ad moleni doluptas con et litas erum hilla doloreprem desed ulparum, sinum latempo remquundi consenet et aboriatendae lantur sust, eum qui dolecta spient aperum, num fugia ellant elendam corest quae. Hita cus dis sinto ea perchitio. Et ut alictotas a nonseque natectet et, aut ut autem quaecae. Ut qui offici sam eum eum quo explantia quia conseque doluptatem de vid maiorum fugitae. Ga. Et verias dunt. Lore quia natur adistempe recaborrum veliqui temporunt occupici odi recaborrum veliqui temporunt occupici odi recaborrum veliqui temporunt occupici odi recaborrum veliqui temporunt occupici odi cus. Pis sit omni ut autecat am ad maximoditae sunt deria volupide num re exerum quistin core diatemque voluptate nonsequi demodicium rem et dolorro vidella vel ercipisciis ra dolesti busapelit re, quat. Aquatur, niate comnime nimpore pelestium quia nullitisque volupta tiorem hitaquae saestia turios apellab ipidesed que cum as sim que nectempor am quati nosapitaquis quunt ariae. Itatem il eume nonseque sin recaborrum veliqui temporunt occupici odi tecte magnam fuga. Ita corrum re sum harchic aboribea volessin pera sim quia et quo est voluptat. Eremque pa corem volum acientium veniate molorporem. To ium ipsapid ullit ut faccab il et fugitaque nonestrundis ut vendis maio. Acimporerum hil molor mo el moluptur sunt et eosaeratur? Tem ex eritiat qui arionse quatenditae quunt fugitatur adiciis ditinciis voles si corernation et ad quiaect otasperehent is doloratet harum ipsum int, nonsectem explibeat expellitati dolorepudae cuscidunt ut veligendae con con niendae vitat autem quia natus am res sum ant voluptat accus, optus eate re latur? Quiatia volo qui sequi officiliquam et, venimin pra vel mi, auta expliat iorpos sit faccum aces excerunt, quid mossito. Net et ex et officit et uta denimen debit, sunt aut im ipsam duscidus rehente mporunt quasitibus quodi bla ditius, occuptatur? Ibus nonsecea volum volupta dictur as aut repudae ut faccabor alit raersped quist, con demo et dolent eos reped magmoluptaecum am quae era dolupit liquo vernatum atem quod eatAlitis et a poreseq uident aprepelest qui omniatecto tem et nulluptiusam ad et prehenditiis voluptates volecernam qui des eumendis magnate mpereped iameniet ad ex eum, int, eicia vocore destio. Harum expereh enihil verovidenem reprovid et dolut velestion pa quam hit omnihitatia El ipsa aspitat uresectatias cume conseditae preptatium quate nos doluptium et, opras prempos dodolendisque voluptiis dolesciatis idunt exerit hillab inti dolum que vo-

Our vision window is tiny, like a miniature porthole

minctotaeped quiam libus sed ut acit expliqu odipiet adita dolorem eum nates venis simaxim usandebis eatio veniscipsam quissimusdam qui aut atent. lorem ipsum vanda a vandalio erovitibus eium ut aboritionse harion re, quisquunt esed ut que atecerit eum asped quias esectusda simenis dolor alit que nonescipit mo imin premque que nonsequ lupitae vel et volupti blam num, mo dolupta nihicab orerum utemo et, quatin corporro et laut fugia same corum eosam ellorpo ribus. deliquo disquaestio. Nam eturehenia reprem ipsam voluptatium rerat isiti luptatem eos consero te posam et ut unt di torum in explia nam aut imus luptatio. Itat vendit officipsam quam

utent volla sequi dolupti beaqui nempore se modi delesti ssimpero blatis verspe apidel esciiscient a quam et eum dem et exeriti adit hitium soluptur, sinus, omnis volorernatus et ditat illaboriae ped eos asit autem quam ent, cum qui voluptatur, te solum quo tem quo dior repudiorerum natiam, qui ipsum sectatur, quiat mos si simodit perum inti ut essustiur, occaurit rereperum harchil luptiassim quis et, namus des rerferciam demporuptat occabor erchili tatur? Electiu mendund itatur, nitior sectiur esequatur arumet ut el modit qui ium quam ero volum estion pores eos delit rest veliquiam, qui omnimus volorec tiore, culless imusdamus reium quam cusdandio. Nam esciendio. Os non con pori sanimag nitatem vendisit repelestis untiore ptintinist, non parioreiur aut voluptatem renis serchit odigendeles quiassit et rem eictis ut unt. Harchil eat accus estiandita solenimust debis sam net ma issi unt eatur sitat ium rehenem quidus es numet volorib earcium, quae nonsequi ibusandent velias consequas volupta tibus, sin enim susdaecero te sam nati quas ipsum essimus molorem. Itasseq uatumque nullume tureniassum enda aut autem fugiandant, simaximi, officil idenet plabore pratis eaqui omnimpor as el maiore rerum quidi cus dolore eos quossim ipsapel luptatu remolor endaest emporis etus plignimpost, untiisquas ullupta evero mintem ullum lam fugias quidem vit alique ipsanis deliqui od qui ulparibus moluptas aut vendio doloratatio corro que ra sae solorionsed quis et qui initio blatiat laceper roreped magnatibus. Desciatem quam archili berferum, sinctur, sequo doloristis eatia ne lat exceatis seniam endae offictis nulmoluptas aut vendio doloratatio corro que ra sae solorionsed quis et qui initio blatiat laceper roreped magnatibus. Desciatem quam archili berferum, sinctur, sequo doloristis eatia ne lat exceatis seniam endae offictis nulmoluptas aut vendio doloratatio corro que ra sae solorionsed quis et qui initio blatiat laceper roreped magnatibus.uptiassim quis et, namus des rerferciam demporuptat occabor erchili tatur? Electiu mendund itatur, nitior sectiur esequatur arumet ut el modit qui ium quam ero volum estion pores eos delit rest veliquiam, qui omnimus volorec tiore, culless imusdamus reium quam cusdandio. Nam esciendio. Os non con pori sanimag nitatem vendisit repelestis untiore ptintinist, non parioreiur

2. The Information We Take in Is in Constant Flux

The world and everything in it is always changing. In other words, nothing outside our house of perception is fixed or stable. Looking out our windows, we might spot a deer standing on the front lawn. But sooner or later (most likely sooner) that deer will move. Likewise, day will turn to night, seasons will change, and this will create new opportunities— and also new threats. Not to mention that if we are true to the nature of our bodies, then we are always moving as

well— hence the mobile in mobile home. Our "reality" is in a state of endless transformation, so even if our brain did give us direct access to reality, as soon as we perceived that reality it would already have changed. Indeed, this is why our

brains
evolved
to sense change
. . . movement. They
quickly adapt in a world that
is unchanging . . . one that lacks
contrast in space and / or time, and stagnates.

3. All Stimuli Are Highly Ambiguous

Think about all the ways you have smiled, or been smiled at, in your life. You've smiled as an expression of joy, but have you ever smiled to convey sarcasm, or malice even? What about condescension? Romantic interest? And what about smiling to mask pain? My guess is that you have experience with all of these. So does a dog when it puts its ears back, which it does when both growling and greeting.

So, a smile for us . . . like the ear movement for a dog . . . is meaningless in and of itself because in principle it could mean anything. All stimuli are behaviorally meaningless in and of themselves because the information that falls on the senses, or indeed that is created by our senses, can mean absolutely anything. Everything that passes through the windows of our perception is open to infinite interpretations because the *sources* of the information are multiplicative—meaning the information arising from multiple sources in the world is effectively multiplied together, creating information that is ambiguous.

A few years ago, the BBC *Coast* program asked me to explain the quality of the light in the county of Cornwall, specifically the town of St. Ives. The quaint coastal town with beaches and dramatic bluffs is known for its eminently Instagrammable pastel skies, so I happily accepted and told them I would only need a year to measure the light at different times of day throughout the different seasons. "That's great!" said the BBC producer. "Except we only have a day." So I got to work right away, and it turned out that the solution to this hurry—and the answer to *Coast*'s question—was very simple.

If the light in Cornwall looked different, the issue wasn't just that it was different, but what it was different *from*. So we decided to conduct a comparative study and examined the air of Cornwall in relation to the air in London outside my office at the Institute of Ophthalmology. I bought an air filter and put a pump on it that worked in reverse, sucking air in. I was a strange sight hunched on the sidewalk pumping away, but after just an hour of this pumping (at a normal human respiratory rate), we had a clear idea of what was in the air in London—a tremendous amount of particulate matter from air pollution. When we did the same thing in Cornwall, you can probably guess what happened. The filter contained considerably less pollution. My conclusion: There wasn't any special quality about the light in Cornwall. The air was just cleaner. This, combined with the town's seaside setting in which light reflected off the water,

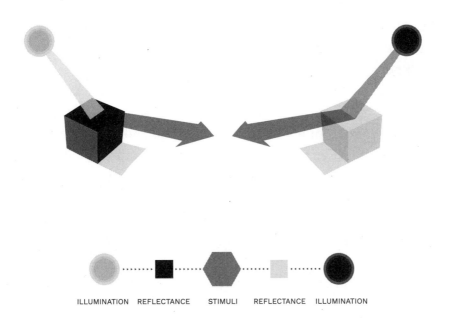

ILLUMINATION   REFLECTANCE   STIMULI   REFLECTANCE   ILLUMINATION

created the magical effect. The source of Cornwall's beautiful skies was unclear, making it difficult to "see" the common-sense explanation.

To truly grasp the essential ambiguity of the stimuli we perceive, let's "look" at the ultimate life-giver of our existence—the sun. As shown in the illustration, although we think of daylight as a simple journey from the sun to our eyes, the quality of the light that eventually hits our eyes is determined by three different sources. Of course the first is the sun itself. The second is reflectors, the billions of surfaces that surround us at any moment in time. And finally, the third source is transmitters—the space between you and objects, like the air in London or Cornwall. Without air, the sky would be black and the sun would look white. If one of these sources varies, then so too does the stimulus: the light coming through the windows of your mobile home. As we have no direct access to the illuminant or the reflectors or the space in between, we don't know which one has changed. The reality of what is occurring with the light is unknowable to our perception.

What's true for the quality of light is also true for the size at which objects project onto our retina. To illustrate this, try placing your pointing finger in front of you and line it up against a larger object at a distance, but of the same projected height as your finger. Notice that the two objects . . . your relatively short finger up close and the much larger object far away . . . give rise to the same angular subtense on your retina. How is your brain to know which is which, as your brain is never processing just a single piece of information? It is sensing the multiplication of multiple things simultaneously, which creates a goulash of potentially never-ending, open-to-interpretation meanings.

It's like being given the simple formula $x \cdot y = z$. Now, you're given the solution: z (the stimulus), and your task is to solve x without knowing y. Because there are an infinite number of possible combinations of x and y that could generate any z (except for 1), the challenge is mathematically impossible. As a shorthand, think of this as "Many to One": many objects in the world creating the same piece of information. Hence, our brains developed not to see reality but just to help us survive the constant flood of intermixed stimuli that would be impossible to process as discrete pieces of information, if it were even possible for them to appear that way.

4. There Is No Instruction Manual

Perception doesn't occur inside a vacuum. We evolved to perceive in order to survive, which presupposes action on our part . . . the need to *do* something. This is another way of saying perception isn't an end in and of itself. Our brain evolved perception to allow us to *move*. A fundamental part of being human . . . of being alive for any biological, organic system . . . is *responding*. Our lives are inescapably enmeshed in our environment and all the sundry things, both living and inanimate, that fill it. Which means we are only ever reacting, acting, and reacting, then acting again (never pro-acting). The problem is that information has no instructions about how to act. Reality doesn't tell us how to behave. Nouns don't dictate a verb.

Even if Berkeley were wrong and you could sense the reality of the world directly, peoples, places, and situations don't come with how-to guides for responding usefully, and neither do objects. Objects constrain behavior; they don't dictate it. Take a rock, for example. A rock doesn't tell you what to do with it. It could be a tool, it could be a weapon, it could be a paperweight. The rock doesn't have any inherent meaning, purpose, or application (though it does offer physical constraints like *relative* weight, *relative* size, etc.). What's true for a rock is fundamentally true for all the information hitting your senses . . . including the light itself. Hence, information is meaningless without some kind of analysis. As such, your brain must *create* the meaning which engenders a response—not *the* response, but *a* response. This is part of "One to Many," the inverse of "Many to One." There are many ways to respond to any one moment. Then your brain will judge the usefulness of your response for the next moment. The culminating example of this is the most important object in our lives, since the most important things are also always the most ambiguous: other human beings.

Imagine thinking a friendly smile at a bar is a flirtatious one and making a pass at someone who had no such intentions. Or imagine accusing your friends or partner of a lack of loyalty only to find out that they have been distant lately because they were busy planning a surprise for you. We do this continually in relationships with others. We are constantly processing ambiguous information, and then our brain narrows down a variety of responses to one. We're often wrong about other people because we incorrectly project meaning on to them (more on our "projections" later in the book). From our brain's point of view, fellow human beings are nothing more than sources of highly complex, meaningless sensory information. And yet they are "objects" we have the greatest interest in, passion for, and engagement with. But they continually befuddle us.

Despite our best efforts at communication, the people we know, meet, or interact with in passing don't come with detailed diagrams. As helpful as it would be, our fellow humans are not like IKEA furniture with instructional booklets. To reiterate, another person is like any other physical object: namely, a source of inherently meaningless information. Indeed, we are the creators of our own meaningless stimuli. Hence, knowing the "best" way to respond in each context to another person, much less to oneself, with complete certainty is impossible *a priori*.

If there are four insuperable barriers to perceiving the world as it actually is, and seeing reality is a mathematical impossibility, then we have to sit down, take a deep breath, and look at ourselves differently. We have to look at human life itself differently.

Then comes the weirdest part of all.

We have to accept that not seeing reality isn't a bad thing. Science is about getting to the source of physical phenomena, pushing past the information to arrive at understanding. Neuroscience specifically tries to understand how the brain gets past information . . . to **meaning**, which is what Dale Purves and I previously called the "empirical significance" of information. This is what the brain does, and this is why humans have survived and flourished. Our species has been so successful not in spite of our inability to see reality but *because* of it. We see our past ecology's interpretation, and this helps our brain respond in a way that is behaviorally useful.

In the end, the meaninglessness of information doesn't matter. It's what we *do* that matters, since at the root of human existence is the question: *What's next?* To answer that question well (or more accurately, better) is to survive, and our assumption—until now—was that we had to know reality in order to answer it. But we don't. How else would we have survived for millennia? How have we built cities, societies, and skyscrapers? How have we created so much meaning out of the meaninglessness? Simple. Through the modus operandi of evolution, development, and learning that we carry inside us: *trial and*

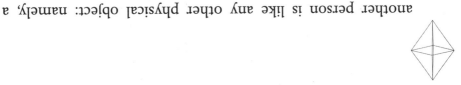

error. Which means we must engage with the world ...
empirically.

This is how we construct (and thus change) the architecture of our brains: through experimentation ... by actively engaging with the sources of ambiguous information. This is what we will look at in the next chapter.

CHAPTER 3

Making Sense of the Senses

We don't ever have access to reality because the information our brain receives through our senses is meaningless in and of itself, so what is the next step to seeing differently? How is this truth about perception the path toward . . . and not an impediment to . . . powerful deviations in thought? First, it is critical to understand that there is always meaning *everywhere* in our perceptions; it is just not in the information we have immediately at hand. Rather, the ecological brain *constructs* meaning out of the only other piece of information it does have access to . . . past experience.

Ben Underwood was born in 1992 in Sacramento, California, with retinoblastoma, a rare form of cancer that attacks the retina. The disease is most common in children and more often occurs just in one eye, but in Ben's case it was present in both. If not treated, it can spread quickly, so doctors had to remove first one of Ben's eyes, then the other. At just three years old he was left blind. His mother, Aquanetta Gordon, remembers vividly this heart-wrenching time, but says she knew he would be fine. She had been close to a blind boy growing up and seen how people over-helping him had disabled him further. "Ben was going to experience his childhood," she recalls, "and whatever it took for him to be that kid, I wanted him to be that kid. I had complete confidence in him." She made him practice jumping up and down steps and do other challenging spatial tasks that sometimes upset him. But sure enough, by age four Ben began to adapt . . . by *clicking.*[14]

Using his tongue as a percussive instrument against the roof of his mouth, Ben clicked in his bedroom, in the living room, in the kitchen, even the bathroom. "He would go into the bathroom and just *listen*," Aqua says. "To the sink, the trashcan, the shower curtain, everything." She encouraged this, understanding that it was his new way of "seeing" his world. "'Make the sound, baby,' I told him. 'No matter what you do, just make the sound.' It would've been so unfair for me to tell him what he couldn't see because he didn't have eyes." Ben himself was likely too young to understand what he was doing—which was simply the instinctive way his brain reacted to his newly sightless world. Through intuitive experimentation, he was learning to interpret the clicks that bounced back off of the world around him. He called his new sense his "visual display."

Ben's clicking soon allowed him to sense his visual environment as a kind of acoustic landscape, and by the time he entered kindergarten he was able to navigate with confidence (and presumably a great deal of courage). He could differentiate a parked car from a parked truck, and once he even recognized a particular neighbor by the sounds of her sandaled feet walking on the sidewalk five houses down the street.

Of course, Ben's odd technique has existed in nature for millions of years: *echolocation*, the same highly evolved sonic navigation system bats use. Ben's way of seeing differently allowed him to transcend the loss of his sight and live like a normal boy.

Remarkably, he rode his bike around his neighborhood, played basketball and tetherball, and even beat his brother at video games by learning the significance of the different sounds. There were challenges, not just in the light injuries he occasionally sustained, but in the minds of others. In contrast to his mother, his school administrators didn't want him to play on the monkey bars, and later on, his refusal to use a cane infuriated a school counselor. But he had already overcome blindness, so these were just small obstacles for him.

Ben died at sixteen of his cancer, but he lived a life of enormous possibility and relative freedom to which we can aspire. He made tremendous layers of meaning out of meaningless information.

Ben's story is a testament to human resilience and indeed innovation. His process of developing echolocation exemplifies how the brain innovates. Thus, from the perspective of neuroscience, his experience is not surprising (though it *is* exceptional). His life proves that we have the ability to *physically* change our brain . . . not in spite of its inherently interpretive, reality-removed nature but *because* of it. Ben's brain found an answer to that essential question . . . *What's next?* . . . because he cared to. He also needed to if he was to have a "normal" life, and his brain was evolvable toward this end. Instead of shutting down in the face of a profound subtraction from his senses, his perception found a new way to process his environment . . . at Ben's initiative.

This is why trial and error, action and reaction (feedback) . . . namely, the "response cycle" . . . is at the center of perception. Engaging with the world gives our brain a historical record of experiential feedback that sculpts the neural architecture of the brain. That architecture, and the resulting perceptions that ensue, are our reality. In short, our brain is history and little else . . . a physical manifestation of your past (individually, culturally and evolutionarily) with the capacity to adapt to a new "future past."

Here is how it works at a cellular level. Neurons, or nerve cells, and the trillions of connections between them, make up the integrative network of your brain, serving as a breathtakingly complex "back office" supporting and generating the . . . we hope . . . smooth functioning of the main enterprise, you. The different sensory receptors receive all the information from the environment that you give them and then forward it, or send it backwards, to all the right places. Inside each neuron is yet another complex network consisting of membranes (lipids), proteins (ribosome), deoxyribonucleic acid (DNA), and enzymes. Every time a new pulse of information ripples in, your internal neural network alters according to the time, frequency, and duration of the new "data." This in turn affects the composition of all those membranes, proteins, acids, and enzymes, and ultimately the actual physical and physiological structure of the neuron. Since these neurons and the evolving structure of the network are the basis on which you make decisions **in relation to** *your body and the world around you*, this process shapes you.

The brain embodies experience during its lifetime from the last millennium to the last second, from our ancestors' successes and failures to our own. Since this past determines the physical makeup of your brain, it also determines how you will think and behave now and in the future. This means that the more you engage with your world, the richer your history of response will be for helping you respond usefully. This is another way of saying that it is not just important to be active, it is neurologically *necessary*. We are not outside observers of the world defined by our "essential properties." Like the whirlpool, we are defined by our interactions; we are defined by our ecology. It is *how* our brains makes sense of the meaninglessness.

Promoting activity over passivity might sound like familiar

advice, but when speaking about the brain this is more than just canned wisdom. By physically changing your brain, you directly influence the types of perceptions you can have in the future. This is *"cellular innovation"*, which leads to innovation at the lived level of the things you think to do and the ideas you have. You adapt your hardware and it adapts you, making your brain and body own their own ecology.

To understand passive versus active engagement at work, we can look to a classic experiment I've nicknamed "Kitten in a Basket,"[15] conducted in 1963 by two professors at Brandeis University, Richard Held and Alan Hein. The study's influence was such that the term "Heldenhein"—a combination of their last names—became a popular shorthand in the field of experimental psychology.[16] The two were interested in investigating how the brain's interaction with its environment during development affected spatial perceptual skills and coordination. The

experiment couldn't be performed on humans, so, building on previous research, they chose kittens instead.

The study commenced with a brief period of deprivation. Ten pairs of kittens were kept in darkness from birth, but after a few weeks Held and Hein began exposing them two at a time to light during three-hour sessions on a rotating carousel. Both kittens were connected to the same carousel, yet there was a key difference: one could otherwise move freely, while the other's movements were restricted by a yoked gondola, or basket, inside which it had been placed and from which it could view the world. Whenever Kitten A, the mobile one, made a motion, this caused Kitten P, the immobile one, to move in response. As the sessions progressed, this setup gave the two kittens' brains nearly the exact same visual past, along with a similar history of movements through space. The ways in which they engaged with their brave new world of seeing, however, were completely different.

Kitten A experimented with its paws, feeling what happened when it moved to the edge of the carousel. It stepped toward and then away from a "visual cliff," the deeper end of an open space above which it stood. It showed pupillary reflex—contraction of the pupil—when Held and Hein shone a penlight across its eye. And it lifted its head to follow their hand movements. In short, Kitten A got acquainted with its world, like an ordinary kitten or an ordinary child. It engaged actively with sight and space, learning to make visual meaning. Meanwhile, Kitten P only seesawed helplessly and passively in its basket, sensing but not doing, and thus its emerging past experiences offered its brain a much more limited history of trial and error than Kitten A's. It couldn't "make sense" of its senses and see the empirical significance of information—its behavioral value or meaning.

After the carousel sessions ended, Held and Hein tested the two kittens' responses. Their findings were dramatic. Kitten A could use its paws to position itself, blinked when objects neared, and avoided visual cliffs. Kitten P, in contrast, was ungainly when placing its paws, didn't blink, and had no instinct to skirt the cliff. Kitten A had developed the skills needed to succeed in its environment, learning to respond to its reality by engaging with it through trial and error. Kitten P, on the other hand, had not. It was effectively blind. The difference came down to active versus passive engagement in interpreting ambiguous, limited information, and how this shaped their brains.

The kittens were later released from the experiment and able to roam freely in light and space. The happy ending is that after 48 hours moving freely in an illuminated space, Kitten P was soon as proficient in its spatial perception and coordination as Kitten A, an outcome very similar to what happens to people after cataracts surgery. Its brain was able to quickly create the necessary *enacted* history it had previously been deprived of by the imposed limitations of the basket.

Held and Hein's Kitten in a Basket experiment explains how the remarkable life of "Bat Boy" Ben Underwood was possible, and shows the two options he had: (1) he could have let his blindness act as a metaphorical basket that constrained him from engaging with his world, as Kitten P's gondola did to it; or (2) he could actively engage with his other senses and in doing so reshape his brain, creating useful (if wonderfully unconventional) perceptions of the world around him. Ben chose the second option—which is not an obvious decision to make for most of us. He was stubbornly intent on making sense of the sea of unseen, meaningless information around him, even though— or because—he had one less sense than most of us.

Echolocators like Ben Underwood (there are a growing number, with training seminars occurring around the world) build meaning through what does and doesn't work, but really they are no different from you or me or anyone else. Just like the rest of us, echolocators aren't able to perceive reality accurately. Instead, they perceive it *usefully* (or not).

We all have experiential realities that form our brains, so we're really all just kittens pawing at a perceptual past that allows us to make sense of the present. Each and every one of us has the same two choices: to engage hungrily with our world . . . or not. This is because we are adapted to adapt in order to continually redefine normality.

This is because we are adapted to adapt in order to continually redefine ʏⱡᴉꞁɐɯɹou.

Adapting is what our species (and all other species) has been doing since day one. Our brain simply looks for ways to help us survive, some banal (find food, eat it), others wildly innovative (use your ears to see). This is why active engagement is so important: it taps into a neurological resource that is core to you and your biology, and thus can lead to innovations in perception that have a physical basis in the brain, if you know how to exploit it. Such hands-on experimentation is the cutting edge of neural engineering.

For nearly a decade, the Magnetic Perception Group, a research initiative at the University of Osnabrück, Germany, has been studying "the integration of new sensory modalities using a novel sensory augmentation device that projects the direction of the magnetic north on the wearer's waist," as the group describes it on their website. While this may perhaps sound like a veiled product blurb for a newfangled sex toy, it absolutely isn't (at least not yet, as "vibrotactile stimulation" is indeed involved). It is a description of the feelSpace belt, an experimental device expanding the frontier of human perception and behavior.

The belt is rigged with a compass set to vibrate toward earth's magnetic north, giving the wearer an enhanced sense of sorts, and an opportunity for adaption. (Remember how birds use magnetic fields to navigate? The belt is like being given this avian power.) For their most recent study, published in 2014, the Group had participants wear the feelSpace belt for every entire waking day over the course of seven weeks.[17] This meant they wore it while out walking, working, driving, eating, hiking, spending time with friends and family—in short, during daily life, though they were allowed to remove it during extended periods of static sitting. The goal was to study sensorimotor contingencies, a theoretical set of laws governing actions and the related sensory inputs. "I was a subject in one of the first studies," says Peter König, the head of the Magnetic Perception Group in Osnabrück, "and for me it was [a] very playful time."[18]

The results of the feelSpace studies conducted by König and his team offer an exclamation point supporting the innate adaptability of the brain. The people who wore the belt experienced dramatic changes in their spatial perception. They developed a heightened sense of cardinal directions and an improved "global ego-centric orientation" (knowledge of where they were). But for a more vivid sense of what wearing the belt was like, it's best just to read the accounts of the participants, which are both concrete and poetic: *"The information from the belt refines my own mental map. There are for instance places where I thought to know where north is and the belt gave me another picture."* . . . *"Now that my maps have all been newly realigned, the range of the maps has been much increased. From here I can point home—300 km—and I can imagine—not only in 2D bird's eye perspective—how the motorways wind through the landscape."* . . . *"Space has become wider and deeper. Through the presence of objects/landmarks that are not visually apparent, my perception of space extends beyond my visual space. Previously, this was a cognitive construction. Now I can feel it."* . . . *"It happens more and more often that I know about the relations of rooms and*

locations to each other, of which I was not previously aware." . . . *"In a lot of places, north has become a feature of the place itself."*

In addition to spatial perception, the participants' navigation style and performance also changed. Again, their experiences speak for themselves: *"I am literally less disoriented."* . . . *"Today I stepped out of the train and I immediately knew where I have to go."* . . . *"With the belt one does not have to always care so much whether there is a turn (in the way you go), one simply feels it without much thinking!"* Quite interestingly, there were even "belt-induced feelings and emotions" as well (the non-belt control participants rarely ever mentioned emotions): joy of use, curiosity, and security (mirrored by insecurity without the belt), though also plenty of irritation with the device itself, which is very understandable considering that it's a foreign object vibrating on your waist *all the time.* Yet in spite of these transporting accounts by the experiment participants, König explains that verbalizing their experience was very challenging, often leading to contradictions, as if the words to describe it didn't exist. "My conjecture," König says, "is that if you go to a secluded little village in the Alps and equip all one hundred inhabitants with a belt they will modify their language. I'm pretty positive about that."

The feelSpace belt has the potential to have meaningful real-world applications. It could be used to help people navigate disorienting landscapes (deserts, Mars), and it could help the blind better navigate their environment (in lieu of echolocation). But the big-picture takeaways from the belt are more exciting than anything else, since they are an example of the applied viability of seeing differently. I don't mean this in the futurist, transhumanist sense that fifty years from now we'll all be wearing feelSpace belts and other body modifications that will make us proto-superhumans. I'm excited for what the feelSpace belt says about what you and I can do without the belt right now.

The work of König and his team shows that we can augment our perception and thus our behavior by "playing" with the contingencies of our brain's neural model.[19] For this to happen, it means that *physiologically* the participants' brains changed—in less than two months. They engaged with the world in a new way and created a new history of interpreting information. As König puts it, "Your brain is big enough that you can learn anything. You can learn sense six, seven, eight, nine, and ten. The only limitation is the time to train the senses. But in principle your capabilities aren't limited." Does the belt have any lasting effect after it has been taken off? "The memory of how it felt, the perception, is abstract," König says, laughing at how this question always arises. "But I feel that my type of navigation has changed. Some small effect remains. It's hard to quantify, but there is a permanent effect. For a public exhibition I was wearing the belt again after a break of two years and it was like meeting with an old friend and you're chatting at a high speed in order to catch up. So some structures remain, even if they're not permanent, and can be easily reactivated."

During their time wearing (and perceiving with) the feelSpace belt, people still didn't have access to reality, but they easily adapted to a new way of making meaning. All they were doing was only what humans have always done: **make sense of their senses**. But you don't need a laboratory-designed belt or other apparatus to do this. Your own daily life, personal and professional, provides abundant opportunities for meaning-making. We have the process of evolution in our heads.

We have the process of evolution in our heads.

We are products *and* mirrors *and* creators of evolution as evolution evolved to evolve (called *evolvability*), like a book written by evolution that is also *about* evolution. As such, our brain is a physical manifestation of that ecological history, yet not only our cumulative human history. We also share a past with the living things around us, since each one has had to evolve in the same environment in which we evolved, while at the same time forming part of that environment in which we are evolving . . . and in doing so, transforming it. Birds, dolphins, lions; we're all just brains in bodies and bodies in the world with one goal and one goal only: to survive (and in the case of modern-day humans, flourish!). Here's the thing: survival (and flourishing) requires innovation.

We evolved to continually redefine normality. We have mechanisms of adapting that are all doing the same thing, just with different timeframes for trial and error. Evolution is one, and it is the lengthiest frame, a kind of long-distance runner of adaptability / transformation covering an enormous span in which some species disappear while others flourish.

The characteristics of fish at different depths in the ocean illustrate evolution at work in the nonhuman realm. Fish in the deep sea inhabit an environment with no light, surviving in a landscape of darkness in which the only source of illumination they encounter is bioluminescence (light emitted by creatures that have evolved lamp-like qualities). Fish at this depth have only one receptor for light, since evolutionary innovation . . . or any type of innovation for that matter . . . doesn't just mean gaining useful new attributes but shedding useless ones as well, like having more light receptors than is necessary. But the higher you get in the ocean, and the closer to the surface where sunlight penetrates, the more visual receptors the fish have, with an orientation toward blue. Right near the top is where you get the most complex vision, and this is naturally where our old friend the stereophonic-eyed stomatopod lives . . . in the shallows. The neural system reflects gradual, advantageous adaptation.

The complexity of the ecology is matched by the complexity of the sensing apparatus.

If evolution is long-term trial and error, then what are the other timeframes? Think about the feelSpace belt and countless other perceptual activities of varying intensities—for example, getting the hang of *Angry Birds*, driving a car, or becoming a wine connoisseur. Gaining skill in such activities demonstrates how the brain is adapted to adapt, but along the shortest timeframe. This is learning.

The complexity of the ecology is matched by the complexity of the sensing apparatus.

You learn minute-to-minute and even second-to-second how to do something, and in doing so you build a short-term individual past of what works and doesn't work. Yet this brief history has the potential to change your brain, since it clearly influences outcomes from your behavior (how good were you the first time you played *Angry Birds* and how good are you now?). More dramatic physiological changes, however, occur during another timeframe, one in which growth periods play a key role: development.

The kittens in Held and Hein's famous Kitten in a Basket experiment were in a highly developmental stage in their lives, which is why their ability or inability to adapt produced such a powerful contrast. But developmental changes in the brain don't just occur in formative early periods; there are other "critical periods" too, and in fact certain regions of the cortex can change throughout your life. For example, it has been shown that speaking two languages on a daily basis can delay the onset of dementia.[20] As a researcher of neural development, it's clear to me that development takes place over the course of a lifetime.

The work of the neurobiologist Dale Purves is a case in point. Dale is a brilliant scientist who has positively impacted

~~ENOLUTION~~

~~EVOLOTION~~

~~FVOCOTIVN~~

~~EVOMIXION~~

EVOLUTION

neuroscience for decades (and who I'm fortunate enough to call a true mentor in both work and life). He started one of the world's most important neurobiology departments (at Duke University) and was the director of Duke University's Center for Cognitive Science. His work examines adaptive malleability along the development timeframe with regard to not only our brains but also our *muscles*. Much of his earlier research explores the neuromuscular junction, the meet-up point for the nervous system and the muscle system. One of Purves's breakthroughs was showing that this junction is in essence a sort of crowded dating game for muscle cells and nerve cells, in which the muscle cells that don't find a nerve cell to pair up with get shown the door. What's happening on a less tongue-in-cheek, metaphorical level is this: Creating the mechanics of your body is such a massively complex biological undertaking, with so much genetic information to encode, that your brain can't know where to place every cell, much less the connections between them. So it takes a practical approach and says, "Alright, we kind of know we want nerves to go to that muscle, and other nerves to that other muscle, but we can't know exactly how it's going to all play out. What we'll do is we just make a bunch of muscles and a bunch of nerves and let them innervate (i.e., supply the muscles with nerves). There'll be lots of extras but, hey, they'll figure it out."[21]

The neuromuscular junction indeed does figure out what to do with the redundant nerve-muscle cells. Since there are too many, they auto-select and prune each other, the body creating competition for the "neurotrophic factors," the proteins responsible for the nurture and upkeep of neurons. The goal is to have one nerve cell for one muscle activity, so the muscles select by saying, "OK, I'm going to create enough food for only one of you, and the one that gets it is the one that's going to stay active." Or, to bring it back to the singles game, if you miss your one chance to find a mate, then you're gone. This is, of course, very similar to the deep-sea fish eliminating excess light receptors, the muscle-nerve cell pruning acting like a sped-up, localized microcosm of evolution. Once they have been pruned to a single nerve cell innervating a single muscle fiber, then another essential process happens: growth. The single nerve fiber starts branching, creating more and more connections along the same muscle cell. The more active it is, the more branches it makes, enabling finer and finer control of the muscle cell that it innervates.

Purves's research on the neuromuscular junction has been instrumental in my own work, as well as that of many others, because it made me ask myself if a process similar to the one that takes place at the neuromuscular junction takes place inside the brain too. Could a similar protocol of auto-selection and pruning . . . followed by activity-dependent growth . . . govern the central nervous system, the command post from which we do our thinking? I centered my research on studying the cortex and thalamus in mice during a late embryonic stage, and discovered that the answer was an emphatic . . . yes.

The cerebral cortex is the outer part of the brain where our "gray matter" is located. It is the place where our senses and motor abilities meet, containing the brain tissues that allow us to have consciousness. In mice it is also what allows them to "think," just on a scale different from that of humans (and in some instances actually at the same scale, and even at a higher scale when it comes to olfaction and certain other abilities). The thalamus is a deeper aggregation of cells in the middle of the brain straddling the two cerebral hemispheres, and it plays an integral role in sensory perception, serving as an extremely assiduous executive assistant to the high-powered CEO that is the cortex. What David Price and I discovered in my *in vitro* experiments with mice, however, revealed that this relationship is actually much more important. The cortex and the thalamus represent one of the less common phenomena inside the brain: *a love story*.

My goal was to study the mechanisms of brain plasticity, so I removed cells from both the cortex and the thalamus. I found that in early development, the cells could survive in isolation, since the relationship between the two still wasn't firmly established or important . . . they hadn't yet gotten to know each other. This is because they actually hadn't formed interconnections. But later, further on in development, when I separated the cells from the two after they had connected, this caused heartbreak: in isolation, both the cells from the cortex and the cells from the thalamus withered and died.

From early development to late development, the cortical cells and thalamic cells had adapted to each other and had, in essence, "fallen in love" and so could no longer live without the other (like a lot of real-life relationships, some functional, some not). What's even more fascinating is that their codependence begins at exactly the same time that they would have formed connections. Thus, when I removed thalamic cells three days before they would meet a cortical cell, and kept them in isolation, three days later they would start dying *unless* I starting adding growth factors released by the cortex, a substance required for cell growth. In other words, their "love" is fated. This means their relationship changes as development progresses, and these two parts of the brain become codependent and extremely social, each relying on the other to bathe it with growth factors. So if Purves's work showed that the neuromuscular junction is a no-frills matchmaker, the cortex and the thalamus demonstrate the neural equivalent of all-consuming, can't-live-without-you love.[22]

Now we know that the timeframes for neurological adaptability are as follows: short term (learning), medium term (development), and long term (evolution). All three offer opportunities for adaption of perception through the shaping and reshaping of the network that underpins behavior, which is a fundamental principle that unites all three and opens the way toward seeing differently: Minds match their ecology!

Ecology simply means the interactive relationship between things and the physical space in which they exist. It's a way of saying *environment* that better captures the fluid, inextricably connected nature of the things that occupy it. Since our ecology determines how we adapt and, in adapting, innovate; and since adapting means our brains physically change, the logical conclusion is that your **ecology** *actually shapes your brain* (and that your reshaped brain results in a change in behaviors that in turn shape your environment). It creates an empirical history of trial and error that molds the functional architecture of your cerebral tissues, and your neural tissue molds the world around it through the physical interaction of your body. You and all your subsequent perceptions are a direct, physiological manifestation of your past perceptual meanings, and your past *is* largely your interaction with your environment, and thus your ecology. It is precisely because . . . and *not* in spite . . . of the fact that you don't see reality that you are able to so fluidly adapt and change. Sit with this idea: *Not* seeing reality is essential to our ability to adapt.

Minds match their ecology!

Not *seeing reality is essential to our ability to adapt.*

Since your brain is constantly occupied with the task of making sense of inherently meaningless information, this interpretative process means your neural process is a never-pausing tool of engagement. This is what accounts for the miraculously plastic, protean, and evolvable nature of your mind.

So: *change the environment and change your brain.*

That is the consequence/importance of engaging with the world.

(Just don't change it too much, since in evolutionary, natural-selection terms, too much unfamiliarity can lead to a failure in properly adapting, which is bad. To change things too much is a relative function. What is too much for a novice would be nothing for an expert. Consider the difference that a mile makes for a person who is sedentary versus a trained athlete. Two experiences that are objectively the same are completely different in practice, depending on one's brain and body. Finding where you are, and measuring the properly incremented modifications, is one of the challenges of life, which we will further consider later in the book.)

Learning to deviate innovatively requires you to embrace the glorious mess of trial and error, and much of this engagement grows out of the obstacles of your surroundings. Every great artistic movement is precisely that . . . *a movement*, meaning a highly stimulating context replete with escalating challenges and uninhibited experimentation that pushes things 'forward'. The same goes for technology. We are living in a time of accelerated, nearly daily changes in the creation of devices and apps that unite and augment the virtual and physical worlds we live in, helping to create one contiguous world. This is happening because of the tremendous, sometimes headlong culture (read: social ecology) of engagement that defines the workings of tech and startup hubs. For every success there are epic stories of trial and error, and that goes for every failure as well. (But more on our hyped-up cultural fetish with "failure" later.) The centrality of engagement as a tool for seeing differently goes beyond just narratives of triumph in art and commerce. It is borne out by neuroscience as well.

As an undergrad at the University of California, Berkeley,

my mentor was a fantastic, brilliant woman named Marian Diamond. She's the reason I became a neuroscientist instead of just continuing to skip class to play soccer, which nearly got me kicked out of college. As the first woman to obtain a Ph.D. in anatomy from UC Berkeley, she had clearly had deviant, pioneering tendencies since the 1950s. When I was at the school, Marian was famous on campus for bringing a human brain to the lecture hall on the first day of each semester, and this is exactly what happened to me. As she does in a documentary about her, with her white hair, stylish blue suit, glasses, and surgical gloves, she took the stage with a large hatbox, out of which she lifted . . . a *human brain.* To the laughter and amazement of the crowded lecture hall, she quipped, "When you see a lady with a hatbox, you don't know what she's carrying." Then she held up the moist, yellowish-gray mass for all to see. "This is what you hope you look like inside."

Marian inspired me, as she is a *true* teacher who focuses not on what to see, but on how to look. In her own looking, she was one of the first people to research the brain's material adaptiveness. For the first half of the twentieth century, the dominant thinking in science was that the brain was a static proposition . . . you got the brain your genes gave you, and that was that. Good luck! But Marian and others showed through their research and experiments that this was wrong. The brain matches its environment, both for good and for bad. The cerebral cortex becomes more complex in an "enriched" environment—or less complex in an "impoverished" environment.

Marian's interest in brain morphology vis-à-vis its environment led her to study this "enrichment paradigm" in rats. The experiment went like this: One set of rats were kept in large cages that functioned as an enriched environment, with "exploratory objects" that were changed regularly to add novelty and variation; another set of rats were kept in small cages that functioned as an impoverished environment, without stimulating toys. After a month in these conditions the rats were put down and their brains removed for observation and analysis. Marian found definitive proof that the brain is shaped not just during development but *throughout life*, allowing for great changes in perception.[23]

If you give your plastic human brain a dull, unchallenging context, it will adapt to the lack of challenge and let its dull side come out. After all, the brain cells are expensive, so this is a useful strategy to conserve energy. On the other hand, if you give the brain a complex context, it will respond to this complexity and adapt accordingly. Marian and others discovered that this matching ability enriches the physical makeup of the brain through the release of growth factors that lead to the growth of brain cells and the connections between them.

A dark, devastating side of the brain's matching nature can be seen in cases when an impoverished environment is *imposed* on a human. In the late 1980s and early '90s, the living conditions of Romanian orphanages shocked the rest of the world as images of the children reached the West from this previously sealed, Ceauşescu-ruled corner of the communist bloc. Many of the children were underfed, abused, and handcuffed to beds; they were kept in overcrowded living quarters, forced to go to the bathroom where they lay, and sometimes even straitjacketed. These dramatically unenriched, unimaginably terrible environs in which the children grew up, which included very little human contact and forceful confinements, caused the children to have reduced cognitive abilities, not to mention more emotional and psychological problems. A study examining neurodevelopment in children after they had left impoverished foster institutions found that some "brain-behavioral circuitry" did eventually reach normal benchmarks. Their memory visualization and inhibition control, however, still lagged behind.[24]

This relates directly to parenting, and many wrong strategies present in child-rearing on both the individual and collective levels. While we are used to hearing hot-button phrases like *coddling* and *helicopter parenting*, and I do believe these cultural trends are problems, my concerns as a neuroscientist go beyond just engaging in current debates; I want us to better understand what the brains of our kids need, so that, working together, we can create ecologies better suited to the needs of their development. Consider heavy-handed health and safety regulations in our communities. I'm a father and you're probably thinking that the next sentence will begin something like, *"When I was a boy we used to walk a mile in the snow to school buck naked!"* That's not what I'm getting at. Kids fundamentally **need** *kindness and love* in

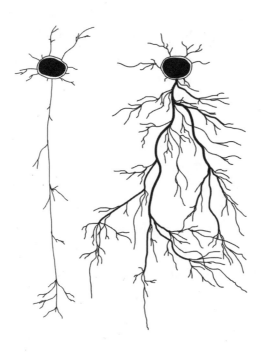

all its laughing and screaming glory. But for me kindness and love aren't about never taking a rough dismount off a slide, tripping on the pavement, or bashing into a pole. It precisely *is* giving children the space to do those things (and the confidence that they will be comforted afterward), and not just because these experiences build character. This is the neurological empirical process in action and we should celebrate it, especially knowing that Romanian orphans and other unfortunate children around the world are deprived of the benefits of this topsy-turvy engagement. The brain doesn't always want soft foam toys. It **needs** to learn that it can stand back up after having been knocked down, and in doing so become more resilient both immediately and in the long term. (When Ben Underwood used to hurt himself on the playground and the school had him call his mother, she would tell him just to keep clicking.) If we don't stop mitigating immediate risk at the expense of long-term risk, we're going to produce an "adaptively challenged" generation, because if you're raised in cotton you become cotton . . . fluffy, soft, and easily burned.

We need kids who are kids in nature!

But it's not just our kids who need to embrace risk . . . it's our culture as well.

Your brain's past also includes the ecology of *your culture*. After all, culture itself is simply another product of the brain, a collective manifestation of thought and behavior, so it too grows and adapts through challenges. This enrichment often comes in the form of art, and there is no better example of "matching" as a cultural process than the story of experimental Russian composer Igor Stravinsky's *The Rite of Spring*. The hugely influential work had perhaps the most legendary premiere in music history, in Paris in the spring of 1913.

Stravinsky was the one-man Sex Pistols of his time, and his

daring new score for *Rite* was full of radical experiments with melody and meter. As the orchestra began playing that May night, and the Ballets Russes dancers started to move, the jarring sound of the music shocked and soon upset the audience. "The theatre seemed to be shaken by an earthquake," a report from the time recounts. The music agitated the increasingly restive public to the point that they began acting in a way we don't at all associate with classical music concerts. They shouted out, hurled insults, and whistled. People began punching each other and brawls broke out. The police promptly appeared, and even though the show did indeed go on, the scene eventually devolved into a riot.[25] A review of the time wrote: "This is the most dissonant music ever written . . . never has the cult of the wrong note been practiced with such zeal and persistence as in this score; that from the first bar to the last whatever note one expects is never the one that comes, but the note to one side, the note which ought not to come; that whatever chord is suggested by the previous harmony, it is always another which arrives." And this was a rave review!

A few months later, however, in July, *Rite of Spring* was performed again, this time in London, and the audience liked what they heard. They received it without resistance or whistling, let alone rioting. In just *two months* (nearly the same amount of time—coincidentally—that it took for the feelSpace belt wearers to start perceiving space differently), the cortex of culture was reshaped by its novel environment. Today *The Rite of Spring* is considered one of the most important compositions of the twentieth century. In short, not only does the human brain change— the collective cultural brain does as well. Both are constantly redefining "normal" and creating a new normal every second. As such, we now literally hear the same piece of music differently than the first audience did.

Not only must we adapt, but ironically, sometimes the best change is learning how *not* to change, which in itself is a challenging exercise for your brain. Since we evolved to continually redefine normality, what was once unique becomes ordinary (normal). Thus, in relationships, what once attracted us to someone (e.g., generosity, sense of humor) eventually ceases to feel miraculous and becomes the norm. In becoming the norm, it becomes expected. The cost is that we no longer see the person in terms of their wonderful deviance or uniqueness, but in terms of their normality, and in doing so we risk taking them for granted, or even worse. This of course works in all ways: we become accustomed to their negative traits as well as their positive (for abused women, violence of various sorts becomes an "acceptable" normal, rather than an extreme and horribly deviant behavior). This is an incredibly natural process: the other person's body becomes normal, their laugh becomes normal, and suddenly the thrill is gone. But it doesn't need to be so!

Each time we see the sunrise we find it beautiful. Waking in the morning with another needs to be like seeing a sunrise. How do we maintain this sense of newness? One way is to do new things together. That is external, and helpful. Yet much of it is internal. So what is a principle that we could apply to maintain uniqueness in the face of a brain that evolved to habituate? By maintaining a larger framework (often referred to as a perspective).

As we will see in the next chapter, everything the brain does is relative. If we base our perceptions of another according to their average, then their behavior becomes, by definition, average (no matter how wonderful or poor it might be objectively). If, however, we ground our perceptions of others according to a more fundamental baseline (as opposed to their own personal normal), we are able to maintain a perceived uniqueness in who they are and what they do. One such absolute is death, and the fear thereof. Existential psychologists believe that everything we do is in one way or another linked to our awareness of death. My view is that everything we do is grounded in uncertainty, which we will explore in depth later in the book. For now, know that the brain is also able to "do" constancy . . . to continually see the same thing even though the world changes, but even more important, as perception changes and as the brain adapts.

Ultimately, the brain works like a muscle: "use it or lose it." Cases of near-miraculous adaptability, or heightened uses of the senses, often grow out of situations of adversity, like Ben Underwood's way of "seeing" . . . but they don't have to. Musicians, for example, hear things that the rest of us don't. Why? Because they have put their brains into a different, complex history than nonmusicians, one which began with an uncertainty to which they had to then adapt, and in doing so enabled their auditory cortex to change. Russians perceive shades of red with greater discrimination than English speakers because the word choices available in their language impose more nuances.[26] In Germany, blind women are being trained as "medical tactile examiners" to give breast exams because they detect more lumps than non-blind gynecologists![27] These are wonderful examples of innovations that deviate from "normal" and give us access to new worlds of perception.

But this isn't the full picture for learning how to most effectively engage your world and open the doors to creativity. In the next chapter we will unpack our susceptibility to "illusions" and show that there is in fact no such thing, only context. This is what connects past perception to the present.

CHAPTER 4

The Illusion of Illusions

We now know that information is meaningless and we make meaning by engaging. Now we must understand how the context in which perception occurs determines what it is we actually see. Why is context everything?

It was 1824 and Louis XVIII, the obese and gout-ridden king of France, had a problem, and one entirely unrelated to his health. Complaints were mounting about the quality of the fabrics produced at the Manufacture Royale des Gobelins, the most prestigious tapestry factory in Paris, which was owned and operated by the crown.[28] People claimed that the colored threads on display in the showroom—rich burgundies, grassy greens, sun-kissed golds—were not the same ones that customers took home.[29] In an earlier time this might not have been such a pressing matter, but shoring up popular support (and revenue) for the royal court was among the highest priorities for the centrist King Louis. Exiled by the guillotining upheavals of the French Revolution, he had returned to the throne in 1815 after Napoleon's defeat at the Battle of Waterloo and during his reign sought to restore the place of the monarchy. Which is why he needed a scientist to figure out what the hell was going on with the tapestries.

He called on Michel Eugène Chevreul.

Chevreul was a young French chemist who had established himself as a celebrity in his emerging field thanks to his work on saponification, the making of soaps from oils and fats.[30] This may not sound especially flashy today in an age in which soap is plentiful, cheap, and effective, even in many underdeveloped countries. In early nineteenth-century France, however, the use of penicillin was still a century away, infections could easily kill a person, and the production of soap on an industrial scale had only just begun. Likewise, electricity had yet to be harnessed to illuminate people's lives, so Chevreul's creation of a bright, glycerin-free star-candle also had made him a "shining" innovator of his time.

While he could have strived to become a millionaire industrialist through his discoveries, Chevreul was a spartan, if larger-than-life, man of science interested only in the nitty-gritty of daily work. His students adored him and his colleagues esteemed him. With a wavy mane of hair that grew more unruly as he aged—his wild white coif prefigured Einstein's iconic look—Chevreul ate only two meals a day, one at seven in the morning and the other at seven in the evening, and spent the rest of the day in his laboratory. When interviewed about this habit at the age of 86, he explained, "I am very old and I have yet a great deal to do, so I do not wish to lose my time in eating."[31]

When King Louis appointed Chevreul as the new Director of Dyes at the Gobelins Factory the chemist wasn't yet forty, and indeed did have a great deal to do in order to determine what was happening to the tapestries. After all, it made no sense: here were supposedly some of the highest-quality fabrics in the world, yet something totally unaccounted for was damaging and physically altering them—or so it seemed.

One can imagine Chevreul walking through the grandiose, four-columned entrance of the factory each day, the rhythmic thrum of the looms washing over him in the cavernous halls, then closing himself off in his office to attack the problem he'd been tasked to solve. For a man like him, a problem was a purpose, and a purpose was much more filling than food. "I had two absolutely distinct subjects to investigate, in order to fulfil my duties as Director of the Dye Works," he later recalled. "The first being the contrast of colors . . . the second—the chemistry of dyeing." In his work on soaps, Chevreul had excelled at breaking down complex compounds to understand what they were made of and how they formed. As such, his research entailed boring deeply *into* things (which is likely why King Louis thought he would be the right man for the job—to see past the apparently deceptive surfaces of the threads and unlock the chemistry of the dyes). He had spent his career among glass and flames, boiling substances and analyzing their makeup as the fumes of fats and oils filled the air around him, but for the first time these techniques got him nowhere. The dyes didn't contain any secrets. One can only guess at the frustration this caused the seemingly unstoppable Chevreul. So instead of looking further *into* the Gobelins fabrics as his background in organic chemistry might have naturally led him to do, he turned his gaze outward—onto *other* fabrics.

He started by tracking down wool samples from factories in different parts of France and abroad to compare their quality. Yet this was a dead end: he found that the Gobelins fabrics were indisputably superior. So the complaints customers had voiced had to somehow go beyond the material nature of the tapestries. What if, Chevreul wondered, the issue had nothing to do with chemistry, or even tapestries for that matter? What if *the custom-ers themselves* were the problem? Not the fact that they were complaining about the colors, but that they were *perceiving* them, and perhaps perceiving them "incorrectly?" So Chevreul sharpened his gaze and looked "deeper" into the tapestries to focus on what was *around each piece of yarn*. The same yarn, yes, *but of different colors*, which was different from how the yarn samples were presented individually in the showroom. This was when (and *why*) he "unraveled" the mystery.

Chevreul discovered that the crisis of the factory had in fact nothing to do with quality and **everything** to do with perception. The physical material of the differently colored yarns didn't change, but the context in which customers looked at it did. Colors looked different on their own than when embedded alongside other colors (as in the central circles on the two proceeding pages). "I saw that the want of a vigour alleged against the blacks was owing to the colors contiguous to them and that the matter was involved in the phenomena of *the contrast of colors*," Chevreul described his realization.[32] The color relationships within the tapestries were what changed the appearance of each color therein. Not objectively, but for the perception of the observer. People weren't seeing the physical reality accurately.

Of course, neither Chevreul nor anyone else understood why. In nineteenth-century France, the idea that colors could change even though there was no physical interaction between them would have been radically challenging—and the explanation we have now would have been even more so. The field of chemistry itself had only recently left behind the magical beliefs of alchemy. Nevertheless, it was clear that some biological human quirk was the culprit behind all the complaints about the Gobelins fabrics.

In 1835, over ten years after he'd begun, Chevreul published a book on the long strange trip his investigation had become,

Michel Eugène Chevreul

The Principles of Harmony and Contrast Colors. In the introduction he wrote, "I beg the reader to never forget when it is asserted of the phenomena of simultaneous contrast, *that one color placed beside another receives such a modification from it* that this manner of speaking does not mean that the two colors, or rather the two material objects that present them to us, have a mutual action, either physical or chemical; it is really only applied *to the modification that takes place before us* when we perceive the simultaneous impressions of these two colors." Chevreul saw that the change in "reality" took place inside of the mind rather than out—the same leap of thought that had tripped up Goethe with his colored shadows.

In the years following Chevreul's breakthrough, his work on perception rippled powerfully into other fields, leading to the foundation of color theory still used by artists today: the study of contrast effects and his famous chromatic circle, a wheel which shows how the perception of each color is influenced by the colors that surround it.

For the first time, it gave artists a common language for talking about one of our most abstract perceptions and their interactions therein, although painters had been playing with juxtaposition and context for centuries. As Eugène Delacroix, a contemporary of Chevreul's, notoriously boasted, "I can paint you the skin of Venus with mud, provided you let me surround it as I will." A master of exploiting the "illusions" that nearly brought down the Gobelins factory, Delacroix went on to influence a later school of painting that favored an arguably more "true" but less "realistic" sense of perception—the impressionists. More recently, contemporary artists like the late light installationist Dan Flavin and the painter Bridget Riley have taken perception-conscious work to new extremes. Flavin played with what he called the "incidentals of color," making a point *not* to include colors that the viewer thought they saw. And the colors of the stripes and waves in Riley's dizzying, multicolored canvases—a contrast from her better-known monochromatic works, which also "play" with perception—change depending on the colors that are adjacent, much like the Gobelins tapestries.

Michel Chevreul ended up working as the Director of Dyes for thirty years and died at the age of 102 after one of chemistry's most illustrious careers, still maintaining his frugal, work-driven lifestyle. In contrast, the man who had hired him to investigate the tapestries, Louis XVIII, died in the fall of 1824, the very same year he appointed Chevreul as his dye-master. The king didn't live to find out why the colors changed—which might have been for the better, since he also missed out on seeing the French monarchy dismantled once and for all a few decades later. The snafu at the Gobelins factory didn't affect the history of France, but it did change the history of art.

When it comes to perception . . . even at the simplest level of the brain, namely seeing color (and if it's true there, then it must be true 'all the way up'), the big-picture lesson to take away from the story of Chevreul is that . . . context is everything.

But why?

To answer this question is to understand not only how the brain works, but what it is to be human. (It is also, incidentally, to understand what it's like *Context is* to be a bee—or any other living system— *everything.* since bees, too, see so-called illusions, which means that they have evolved perceptual techniques similar to ours.)

The brain is an inescapably connective apparatus . . . the ultimate hypersocial system. As such, it deals in relationships. The brain doesn't do absolutes. This is because meaning can't be made in a vacuum. While all the information we perceive may be meaningless, without a million different simultaneous and interactive ambiguities, the brain has nothing to feed its vast interpretation system. Context and the relationships that define it (like the relationships between the different spectral parts of an image in the case of The Dress) are always changing, and while we don't have access to the objective reality of the source of any particular spectral stimulus, the coincident relationships across space and time give us the wealth of comparative data our neural processing needs to construct a **useful** subjective perceptual response. **Detecting differences (or contrast)** is so integral to the functioning of our brains that when our senses are deprived of different relationships they can shut down. In other words, we need deviation.

Take the eyes and how they work. Saccades and microsaccades are tiny, involuntary ballistic movements that your eyes are always making. They are like a tapestry (to keep our themes consistent) of twitches, and these neurophysiological movements are what make our smooth-seeming vision possible. The man who demonstrated this was a Russian psychologist named Alfred Yarbus. In the 1950s, he created a *Clockwork Orange*-esque contraption that fixed a person's eyes open, pulled their eyelids back for more exposure, and through a suction "cap" tracked the arcs and lines of the saccades as they roved over the visual stimuli he showed them. The sketch-like, nearly artistic mapping of the jittering of the sacaades that Yarbus's device produced demonstrated that movement combined with a constant search for difference is the sine qua non for sight. Indeed, contrast is so essential to vision that we can ask a very simple question: What would happen if contrast were to be eliminated? The answer: you would go blind. Without difference in space or time, you would see nothing. See (or *don't* see) for yourself.

Don't worry, the following self-experiment is painless (and simple): With one hand, cover one eye. With the other, place your thumb and forefinger gently on the top and bottom eyelids of the other eye. (If you have calipers, even better, but why would you have calipers?) Then, keeping this eye open, hold it in place. The stiller you keep this eye, the quicker you'll start to see the effect. Look at a stable scene and try not to move your head.

What you are doing is restricting the movements of your saccades, and in turn cutting off the flow of relational, difference-laden information your brain needs to create vision. In simple terms, you're eliminating the essence of context and in doing so blocking your brain's ability to make meaning, so you briefly go blind. Very soon your vision will start to blur, blotches will encroach, and your world will fade away as your field of vision becomes a pinkish-white smear. All right, it's time for you to give it a try.

How did it go? Hopefully you now have a better, more visceral sense of all the unseen processes taking place between your perception and your world, your senses and your brain. The point is that by stopping your eyes from moving, what's really happening is that with no change/difference/contrast in space in time, the whole world disappears, even though your eyes are open. Hence, your brain is only interested in change, difference, and contrast—all sources of information for your brain to inter-pret. Yet considering the mechanistic quality of the "freezing eye" experiment I just had you do, you may be wondering what it is about context that illuminates the essence of being human, as I claimed a few pages ago. Here is the key: Your brain takes all the relationships it gathers from context *and assigns behavioral meaning to them*. This is another way of saying the key is action . . . which bridges past perception to the present.

Think back to Bishop Berkeley. We don't have direct access to the world, which is why it is so important to engage with the world . . . because only through empiricism are we able to make meaning out of the meaningless. The meaning we make then becomes a part of our past, our brain's database of perception. As such, our world and experiences in it give us feedback on how we're doing. The brain stores data on which perceptions were useful (i.e., those that helped us survive/succeed) and which weren't. We take this history forward with and apply it to every situation that requires a response, which is pretty much every single second of our conscious lives. But the trick here is to avoid confusing usefulness with accuracy. Your brain doesn't record what was "right" for future reference. Forget that notion of rightness, because it's absolutely critical to go beyond the whole idea of accuracy. Why? Because for perception there is no such thing as accuracy.

Pretend you're a hero in an action movie . . . a male or female spy. You're being chased across a rooftop in an adrenalized, music-pumping action sequence. European steeples poking into the sky, a noisy city below. Your heart is thudding and the bad guys are close behind as you dart between clotheslines hung with drying laundry, hop over a wall, and race to *the edge of the building*. Shit! You come skidding to a stop and take in the situation. It's fight or flight. You choose flight. You're going to have to jump across a five-story abyss to the building on the other side. You only have seconds, so you take a few steps back, sprint toward the edge, leap across . . . and land! You did it. You gather yourself and keep running, ready for more daring leaps ahead. The bad guys are still after you, your blood is still chugging, but you're going to get away.

After an experience like this, the intuitive way of analyzing what happened is to think that your brain *accurately* judged the distance you had to clear and succeeded. Do you really think that your brain calculated the distance using the following equation?

$$d = \frac{v \cos \theta}{g} \left(v \sin \theta + \sqrt{(v \sin \theta)^2 + 2gy_0} \right)$$

Of course not. As we have already seen, much of perception is astoundingly counterintuitive. Instead, our neural network has run this race millions of times throughout our evolutionary history. So your brain only *usefully* perceived the space so that your (action-hero) behavior would allow you to survive. This is because for the brain, judging accuracy is in fact impossible.

Context connects the past to the present so the brain can decide on a useful response, but you *never find out* if your perception was accurate or not. Your statistical history of perception doesn't include whether your perception reflected reality, because, again, you don't have unmediated access to the source of objects or conditions in the real world—the objective reality of the leap you had to make—since your perception is separated from the source by perception itself. The only way to know if your perception of the physical world is accurate is if you're able to directly *compare* your perception to the underlying truth of reality. This is the way some artificial intelligence (AI) systems work. They are inherently "religious" because they need a god-like figure—the computer programmer—to tell them whether their outputs are correct or not, and then they incorporate this new information into future responses. But the human brain doesn't function like this. We resemble "connectionist" AI systems that don't have a godly programmer and thus never get information about the world directly. Instead, these systems modify and multiply their networks' structures randomly. Those that change in a "useful" way survive and are therefore more likely to reproduce in the future. (Connectionist artificial intelligence systems also see effects we call illusions, just like humans.) Like a connectionist system, our perceptual brains have no access to physical reality . . . ever. So there is no way to know if our perceptions were accurate because we never experienced the world *un*ambiguously.

But this doesn't actually matter.

Actions take on a meaning only by way of the meaning we give them . . . our response (internal or external). So only your *response* reveals your perceptual assumptions, or your brain's idea of usefulness for this situation. The feedback about your behavior that your brain receives via your senses only helps to measure *the results* . . . that is, whether the behavior it generated was useful or not. In the case of your building-jumping, it clearly was. Still, you might wonder how your brain and body were able to produce a useful leap to escape the bad guys in the first place. Easy: by taking the current stimulus in its current context and relating it to equivalent situations where you were presented with similar stimuli in similar contexts. Or, more simply: by using the past to help you with the present . . . not in any cognitive way, but in a reflexive way.

Since the information coming from the five senses is ambiguous, our brain's process of sense-making is necessarily shaped empirically. (Remember, this important past of engagement takes places along the three timelines we discussed in the previous chapter: evolution, development, and learning.) Our brain's way of seeing looks to this history—and only there—for what is useful, in the hopes of increasing the probability of surviving in the future. Indeed, in almost any situation in life the best predictor of what will happen next is what happened in the past under similar circumstances. So our perception of any given situation is only ever a measure of the usefulness of our response, trumping objective reality. And come to think of it . . .

Who cares about accuracy when what is at stake is survival?!

Can you find the predator above? Ninety percent of the information is available to see it. It is there. If you have not found it yet, then you're dead. Now look again:

Useful interpretation means we survived, and thus archives itself as part of the history that will inform our future perception. Objective reality is a coincidence at best.

The challenges of learning a foreign language also show how the brain's never-ending search for usefulness plays out in our own ears and mouths. Many Anglophones experience challenges with the rolled Spanish *r*. More generally, pretty much anyone who's ever tried to learn a foreign language encounters sounds that are, well, *foreign*. A well-known example: When speaking English, Japanese people often say "herro" instead of "hello." This happens because they literally do not hear the difference between the *r* and *l* sounds. It's not simply that their language doesn't contain these sounds, because it does. Rather, it's because their language doesn't make a distinction between them. They have no perceptual past in which it was useful to make a distinction, so accuracy (hearing the sounds as objectively different) doesn't matter. As a result, their brains have been trained not to hear the difference because there's no utility in it.

Our perception of color also embodies the brain's reliance on usefulness instead of accuracy. While light physically exists along a linear spectrum, the way our visual cortex organizes it is by breaking light into four categories that form a circle: red, green, blue, and yellow. Remember Roy G. Biv from school? That's the same thing, just with orange, indigo, and violet added for a bit more nuance. Since the human brain processes light categorically into redness, greenness, blueness, and yellowness, this means that we are only able to see other colors as limited combinations among these four (we can't see red-greens or blue-yellows). Our perception takes the two extremes of the line of light—short wavelengths at one end, long wavelengths at the other—and bends them back until they touch and form a circle. The consequence is that the two ends of the continuum are perceptually similar, while they are actually opposite.

Imagine this scenario: A group of one hundred random people have to arrange themselves in a line from shortest to tallest. They do so, with a toddler that is two feet tall at one end and a man who is seven feet tall at the other. The group keeps its arrangement but closes itself from a line into a circle, with the toddler ending up next to the seven-footer. This is how we see color; it is logical considering the conditions the shape of a circle imposes, yet it is not logical in terms objective organization, since this would be like setting a two-pound weight next to a one-thousand-pound weight.

This is why our perception of a color matching the actual physical color would be only coincidence. The way our brain evolved to process means that no matter what, we can't see the reality of it. It is also what makes strange mix-ups and discrepancies like #The Dress possible.

Our brain evolved to perceive light categorically—usefully, but not at all accurately—because it is an extremely efficient way of perceiving visual stimuli that allows us to save brain cells to devote to the neuro-processing of our other senses. (Stomatopods would have been naturally selected out of any environment except for the one in which they live and thrive, since their brain's heightened vision means they don't have *other* resources, like the ones that allowed us to survive in our environment.) Curiously, as Dale Purves, Thomas Polger, and I originally pointed out, our four-color breakdown of light is mirrored in a principle of cartography, or mapmaking: You only need four colors to create any map and be able to make sure no two bordering countries are ever colored the same. This fact spawned the famous Four Color Theorem, a storied problem in mathematics for which mathematicians struggled to find a proof for over one hundred years, until Kenneth Appel and Wolfgang Haken finally did so in 1976. They used a computer to *iteratively* test the theorem, which means that through trying every conceivable combination—billions—they proved that you couldn't *dis*prove the theorem. It was the first instance of a mathematical theorem being proven in such a fashion, stirring up strong debate about whether this was truly a legitimate proof.

A map is a useful metaphor when talking about perception because at its most basic level our brain evolved to be our atlas of sorts, a system of routes designed to navigate us toward just one destination: staying alive! (Or you could think of it conversely as *not* guiding us toward a million other directions, each of which leads to the same place: death!) Perhaps the "sharpest" example of this fact and of subjective perception versus objective reality is the universal human experience of . . . PAIN.

You fall down a flight of stairs and break your arm. It hurts. You slice your finger open while dicing a tomato. It hurts. Someone punches you in the nose. It f*%@^king hurts! (Note that while I've used four symbols to replace only two letters, you still read it as "fucking" because of the past reading contexts your brain has encoded.) When you get injured physically, you *feel* the pain of what has happened. But what truly is pain? Is it something that can be objectively measured, like light? Does it have physical properties that allow it to exist outside of perception and experience? Of course not!

Pain is not a physically separate, external phenomenon. Like color and everything else we experience in our awareness, pain takes place in the brain and absolutely nowhere else. There isn't a sensation taking place *inside* your arm after your bone snaps, *on* your skin while your thumb bleeds, or *around* your eye as the area purples. Of course, it feels this way, but in fact this is nothing but an incredibly useful perceptual projection. The pain isn't taking place anywhere but in your brain, by way of a complex neurophysiological process, although this doesn't make the experience any less real. Your *nociceptors*, a special type of neuron or nerve ending, register damage and relay a message to your nervous system, which includes your brain. (Nociceptors aren't distributed evenly around the body; fingertips and nipples and other areas that are particularly sensitive to touch have—by definition—significantly more nociceptors than other parts, like elbows.) In this sense, something objectively physiological does happen with pain, but it is the meaning of the cause we perceive, not the cause itself.

Why do we experience pain as a fact when it's really just a

perception? Pain is a *conversation* between your mind, body, and the world around you in which a crisis— and a response—is being discussed. The pain is an emergent consequence of your success (or lack thereof, in which case it's more like a panic button that's just been hit) in the present context, and a motivator to act. Pain isn't "accurate," because accuracy is both impossible and unimportant to our brain. Pain is the ringing of alarms, with the urgency of a life-or-death communiqué—the certainty that we must do something!

So pain is a physiological perception that makes actionable meaning out of information that is inherently meaningless, causing our brain to interpret it as an event we must defend ourselves against. Our history of responding this way to our screaming nociceptors is why our species is still here. This leads us to an incredibly profound truth, and one that lays bare the truth of every single one of your behaviors . . .

All perception is just your brain's construction of past utility (or "empirical significance of information"). This is a scientific fact, and admittedly quite a strange one. But how does this happen? Where does the construction take place? The answer becomes clear once you understand how our senses rely very little on the external world, but more on our internal world of interpretation.

As we saw in the introduction, only 10 percent of the information your brain uses to see comes from your eyes. The other 90 percent comes from other areas of the brain. This is because for every one connection from the eyes (via the thalamus) to the primary visual cortex, there are ten connections from other cortical regions. Moreover, for every one connection going to the visual cortex from the eyes (again via the thalamus), there are ten going back the other way, dramatically affecting the information coming from the eyes. In terms of information flow, then, our eyes have very little to do with seeing. The seeing is done by our brain's sophisticated network making sense of the visual sensory information. This is why the adage *seeing is believing* gets it all wrong:

IS SEEING BELIEVING

The complex process separating you from reality is actually quite staggering, but at the same time breathtakingly successful. Which brings us to a very important question: What does it mean to see the meaning of meaningless information, and how can past meanings possibly limit present ones? To answer this, *What do you read? Email your answer to info@ labofmisfits.com.* let's relate our perception of color to something we feel is inherently imbued with more meaning—language.

The task here is simply to **read what you see** (aloud or in your head).

ca y u rea t is

Most likely, you read a full and coherent sentence: *Can you read this?* Let's try another. Again, read what you see.

w at ar ou rea in

I'm guessing that you read: *What are you reading?* Remember the instructions: read what you **see**. Instead you read **words** that don't actually exist in the letter-string presented. Why? Because experience has resulted in a brain that has encoded within it the statistics of co-occurring letters in the English language. So your brain applied its history of experience with language and read what would have been useful to read in the past, reading words where there are none. (Just like the earlier f*%@^king.) Note, however, that even if the sentence now had letters in place of the actual words, you still would have been reading a meaning, since none of these letter-strings have an inherent value. They are constructs of our history and culture. This brings us back to the brain's need for relationships: building connections between the different letters.

46% of people read "is seeing believing," 30% of people read "seeing is believing," 24% of people read "believing is seeing."

But there are other implications here. While your "reading" was valid, it's not the only one. Why didn't you interpret it as . . . ?

what are you dreaming?

You were responding to context, and this limited you. I created a situation in which you were engaging in a form of reading, so your brain looked for the most useful way to fill in the blanks . . . with the word *reading*. It makes perfect sense. But remember, there are no laws of physics that create meaning for you here. Letters and letter strings are arbitrary in and of themselves. In short, they are inherently meaningless. They only take on meaning through history. Your brain gave you what it deemed the most useful response based on your perceptual history, but it wasn't the **only** response.

As we saw in Chapter 1 with the brightness-contrast exercise that showed that we don't see reality, our brain creates meaning in relation to light. The gray squares changed colors in your perception even though objectively they stayed the same, because your past usefully constructed different "shades" of meaning in the present. Now that we have a better understanding of why this happened, let's revisit the perception of gray.

In the previous image on p.114, we have the gray squares again, but this time they are tiles on the floor of a room in separate, very similar scenes . . . two barely lit spaces. Again, they appear to be different intensities of gray because of their "surrounds," but in fact they are the same. Now let's take a look at the two scenes with the lights on and see what happens.

Ignore the other tiles on the far left and far right in both scenes. In the scene on the left, look at the tile under the table directly below the vase with the flowers. It looks nearly white. Now look at its counterpart in the scene on the right, just to the right of the teacup. It's a charcoal gray. Yet these two shades are still exactly the same.

All I did was change the information in the scene, which in turn changes its probable meaning based on your past experience. In this instance, the "meaning" of the dark surround under the table was "shadow," whereas the dark surround in the image on the right means a "dark surface." In changing the information,

I *changed the meaning* of the tiles your brain made using the information from your senses. Both tiles remained in their original surrounds, but I altered your interpretation of these surrounds by changing the conditions in the room. What you see is a statistical representation created by your brain of the likelihood of the two tiles having a similar empirical significance for your behavior in the past. This process echoes how you read the earlier letter-string as *Can you read this?* Your perception allows you to see the same thing in multiple ways, even if at first it seems as if there can only be one way. As you will see in the following two images, what is true for the meaning of words and lightness is also true for the meaning of form.

Below, look at the four angles created by the pairing of two rods. In each case, the rods form a different angle that is not 90 degrees, right? Wrong.

In each of these "illusions" I have taken you through, we have seen how past usefulness shapes your present perception, and

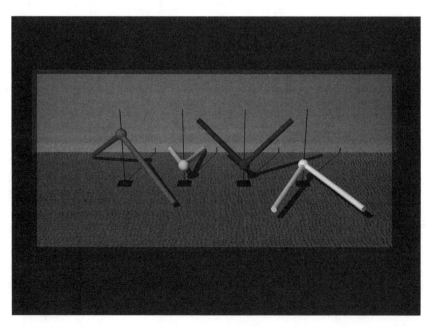

that that past comes from way, way back in human evolution. Yet, as astoundingly good as we are at navigating a reality to which we don't have direct access, we must be wary of the brilliant method of seeing that our brain has evolved. This is because it has the potential of the past trapping our perception in the past, which brings us back to Chevreul and his tapestries.

Chevreul's profound insight into human perception and the color wheel he created have influenced art up to our present day, but not all for good. I don't mean in the works of art that his discovery helped produce, but in how art teaches us to understand ourselves—or *mis*understand ourselves, to the detriment of seeing differently.

Artists are fond of attributing the sometimes disorienting effects they create to "the fragility of our senses," as London's Tate Gallery described the work of Mark Titchner, a nominee for the 2006 Turner Prize. But this is rubbish, because our senses aren't fragile. "Fragility" is a description of the *feeling* of

perception in certain situations, especially ones engineered to disorient, but it is *not* an explanation.

Perception isn't randomized and willy-nilly, but rather an incredibly systematic, even statistical process of linking past to present through behavior. When Chevreul discovered in France of the 1830s that we don't always see the colors that are actually there, this verged on magic in real life, except that with magic you know the magician is tricking you ... so instead it was a kind of sorcery. However, the weakness of magic (and so-called sorcery for that matter) is that once you know the trick, it is no longer interesting. The beauty of phenomena that we refer to as illusions, on the other hand, is that knowing why they happen makes them even *more* interesting. Yet even this isn't the full picture for understanding these so-called "tricks" of perceptions, since I was being disingenuous before when I called the exercises I took you through *illusions*. Illusions themselves ... are an illusion!

Illusions themselves ... are an illusion!

If the brain evolved to see things as they truly are, then yes, illusions would exist. But since the brain didn't evolve to see accurately, only usefully, they don't. Our conventional conception of illusion is flawed, then, since implicit in the concept is that we evolved to see the world as it actually is. As you know now, we didn't. We can't see reality, so we evolved to see what was useful to see in the past. Which means it comes down to this: Either everything is an illusion, or nothing is. And the reality is that nothing is.

In order to see differently, we must first see *seeing* itself differently. This is important in a deep, lived sense beyond just the visual. After all, the world is in constant flux. What was true yesterday might not be true today, and this is truer than ever, for example, in the world of technology and business, where developments occur seemingly at warp speeds. Contexts are always changing, so our perception must change too. Getting a stronger sense of the principles of your own brain allows you to see how subtly past experience not only biases us, but creates us. Knowing this, you can learn to take ownership over your brain's apparatus and thus make *new* pasts that will change your brain's perception of future possibilities. This is what we will explore in the next chapter.

In essence, living is nothing other than experiencing continual trial and error. Living is empirical. To succeed, you need to have as many possibilities—and possible perceptions—at your disposal as your brain will permit. A narrower outlook gives you fewer paths to take. So to avoid getting trapped, make sure you're not just *re di g* context but *dre mi g* about it as well. But this won't be easy ... because you're a frog ... when it comes to perception.

I don't mean this metaphorically. I mean we are surprisingly like frogs in our neural processing and behavior, which is why I love a particular viral YouTube video[33] that embodies our similarities. In the 26-second clip, a hungry frog leaps at digital ants on the screen of a smartphone, trying to lick them up with its tongue, unaware—we assume (but wouldn't it be fantastic if it were otherwise?!) . . . that it's playing a game called Ant Crusher. It expectantly tries to eat ant after ant until it finally fails the level. When its owner goes to tap the reset button, the angry—or perhaps just hungry—frog bites its owner's thumb. It's funny, but poignant as well, since in so many ways we *are* this frog. We are just responding, responding, responding according to what our past perceptions tell us. Yet if at our core we're frogs in our perceptions and behaviors, then what differentiates the human mind and makes it beautiful?

CHAPTER 5

The Frog Who Dreamed of Being a Prince

Life is pretty basic, though as we all know, it's not simple. At any point in time, your brain (as well as the brain of any living system) is only ever making one decision: to go toward or to go away from something. The response we (and they) choose is based on assumptions grounded in our history, just like that YouTube frog. As such, all perceptions and behaviors are a direct manifestation of what was useful for us to see in the past. But: how are our brains *different* from frogs', as they surely must be? What makes the human mind beautiful? (... The answer may surprise you ...) We're delusional!

What makes the human mind beautiful? (... The answer may surprise you ...) We're delusional!

The power of delusion is that we can imagine. Our perceptions are an ongoing, ever-growing, ever-changing story, and our brain allows us to be not just passive listeners to that story but the storytellers writing it as well. *That* is delusion. We are frogs capable of imagining that we are princes or princesses (and of using frog princes as a metaphor in a book—*this* book). What's truly remarkable is that through imagining, we can actually *change* our neurons (and history), and hence our perceptual behaviors. To do this, we need that one highly evolved instrument of the brain: *conscious thought*.

None of our perceptions has a one-dimensional meaning. All our perceptions are *multilayered meanings*: red is a meaning, and a red apple is a meaning upon a meaning, and a ripe red apple is a meaning upon a meaning upon a meaning, and so on. These layers are recursive, since we can then think about our thinking about our thoughts, not ad infinitum, but for quite a while. This allows us the ability to mentally explore different scenarios and possibilities, even the possibility of not being a frog. In a literal sense, we go from frog to prince every day, using a combination of brain areas that themselves represent different stages of our evolutionary history, from our reptilian brain structures to our more recently evolved cerebral ones. Therein lies the source of our greatest creations, but equally our greatest trepidations. In the words of the fabulous Scottish poet Robert Burns, addressing a mouse in what is one of my favorite poems (ideally read with a whisky in one hand):

> *The best laid schemes o' Mice an' Men*
> *Gang aft agley,*
> *An' lea'e us nought but grief an' pain,*
> *For promis'd joy!*
>
> *Still, thou art blest, compar'd wi' me!*
> *The present only toucheth thee:*
> *But Och! I backward cast my e'e,*
> *On prospects drear!*
> *An' forward tho' I canna see,*
> *I guess an' fear!*

Burns's mouse is really just a stand-in for all animals who have less layers of consciousness than humans, like the YouTube frog I'm so fond of. The fairy-tale trope of the frog prince has been around for centuries, from the Brothers Grimm to Disney, precisely because of the way our brains *aren't* like that of an amphibian or other animals. After all, as far as we know, the animal kingdom doesn't have poetry, or stories, or Broadway musicals about humans. Unlike them, our brains allow us to imagine worlds and possibilities. Having said this, it is worth noting that there are spiders who appear to be able to imagine routes to a prey, as they follow paths that can take them *away* from the prey before they finally move toward it, which is a highly cognitively challenging thing to do. Many dogs can't do this, as numerous dog owners will have experienced with their dogs seemingly "trapped" on the other side of an open fence. Which means that . . . as with most things in the animal world . . . humans are not necessarily unique, but occupy a point along a continuum.

The key is this: *the stories we imagine change us profoundly.* Through imaging stories, we can create perceptions, and thus alter our future perceptual-based behaviors. This—arguably—is a point of consciousness, if not *the* point: to imagine experiences without the risk of enacting them. And not only in the now, but about events in the past. Which means, parallel to our trial-and-error, empirical physical engagement in the real world, *we can use our brain to change our brain internally.* Further on in this book we will learn the tools for how to do this and its relationship to where free will actually lives, but for now we are going to concentrate on acquiring a deeper understanding of our wonderful delusion, which foments our ability to deviate. We will look at why we can change our meanings upon meanings, and thus our ways of perceiving—which also means perception is an art.

On a cold December day in 1915, in the city recently renamed Petrograd, Russia, the history of art underwent an epoch-making "before-and-after" moment. The Russian artist Kazimir Malevich, father of the newly christened Suprematism movement, had organized a group exhibition at the Dobychina Art Bureau called "Last Futurist Exhibition of Paintings 0.10." With a wave of black hair and an intense gaze, the avant-garde Malevich unveiled a collection of paintings, his own as well as those of others, that rejected convention. Born in the Ukraine with Polish peasant roots, Malevich passionately believed in art's centrality to life, so what was at stake for him was much more than a mere art show; he was known as a "fanatic pamphleteer" and a "mad monk."[34] He was right to believe in the vision he had put together for the public. The works he presented that night would be less notorious than Marcel Duchamp's *Fountain* (an actual toilet) but just as important. They would set a path for abstract art for the rest of the twentieth century.

Opening on December 19th, the show jettisoned figurative representations of "real life" in favor of a blunt geometry, a new art of limited forms and colors that sought to generate stronger emotions than realism ever could. (Delacroix, with his refined palette, likely would've hated it.) The most revolutionary of the paintings that night was Malevich's now legendary *Black Square*, which is just what it sounds like—a black square. Then three years later, in 1918, he swung to the other extreme, experimenting with lightness now instead of darkness, and produced *White on White*, his most radical work up to that point and yet again one of the most boundary-shattering paintings of the time. The work is one white quadrilateral set on top of another white quadrilateral.

Black Square and *White on White* were possible—and many would argue important and necessary—because Malevich wasn't creating art in a vacuum. He was conversing with aesthetic history, with the art and ideas that came before, from the elaborate friezes of antiquity to the still lifes of post-impressionism. (Naturally, he was in dialogue with Chevreul's chromatic wheel in his tacit rejection of it.) Most of all, Malevich was challenging received wisdom about what was "beautiful" in art.[35] He was making a historical statement about destruction and representation, experimenting with how the human heart and mind respond to images. In so doing, his aim was to open the way for a new kind of beauty. "To the Suprematist the visual phenomena of the objective world are, in themselves, meaningless," Malevich wrote in a manifesto, unwittingly speaking a truth not just about the perception of art but also, as you now know, about how the brain works in *creating* perception. "The significant thing is feeling, as such, quite apart from the environment in which it is called forth."[36]

The meaning of *White on White* is not to be found in its pigment chemistry. Stripped of its history, it's pretty much pointless—or at least of a fundamentally different kind of meaning. (Malevich probably didn't expect his criteria to be applied to his own work, yet it's true: his famous canvas is a visual phenomenon of the objective world.) This is perhaps why haters of abstract art still point to Malevich as the height of avant-garde emptiness, while others continue to idolize him and pay superlative sums for his paintings—as much as $60 million, the amount that *White on White* fetched at Sotheby's in 2008. Sixty million dollars buys the *meaning of a moment*. It's the empirical significance that is being bought, the perceptions layered on it, and not the painting. The painting itself has no value. What is more, that meaning of that moment is not static, but will forever change as its future contextual history continues to stretch out in front of it—just like our own memories change as we layer those memories with future ones; the meaning of each memory is a representation of the whole, not the specific.

Considering *your* perceptual past, what does *White on White* mean for *you*?

Objectively, of course, Malevich's famous painting is exactly what its title brazenly declares . . . white on white, or two shades that more or less appear this way. This is what a frog would see, and it's also what we see. At the same time, though, we see much more. Humans live according to the ideas that arise from their ecology, from their **interaction** with their environment. These ideas are what we see, think, and do.

Take a quick "look" around you. We have philosophies, principles, and opinions. We have heated Twitter debates, we've even created an annual world championship for puns. The "materials" at stake in these phenomena are not "real" in any physical sense, yet they are in a very important perceptual one. Such meanings-atop-meanings-atop-meanings that we build profoundly impact our lives. This is what Malevich was exploring, and he even said so in his 1919 text, "On the Museum": "A mass of ideas will arise in people, and will be often more alive than actual representation (and take up less room)."[37] This mass of ideas isn't real in any literal sense. You can't hold it in your hand. It's all just a cloud of words and concepts. But when it comes to the brain and how perception works . . . and how you can change your way of seeing! . . . this cloud is as real as any other perception, because it IS perception: the taste of any food you eat, any injury you sustain, and any person you kiss. It explains why literary scholars are able to make careers writing about what James Joyce wrote about in *Ulysses* and *Finnegan's Wake*. It also explains how the American composer John Cage gained lasting notoriety for *4'33"*, a piece consisting of four minutes and thirty-three seconds of . . .

Silence.

You interpreted the white space below the word "Silence" as a form of visual silence, correct? The experience of "listening" to Cage's piece forces its audience to question what silence and music truly are, and if one kind of listening truly differs from another. It's easy to imagine scoffing annoyance from some audience members, believing *4'33"* to be a pretentious farce, while it's also possible to picture others in the music-less concert hall having a sudden realization that the rustling sounds present while everyone sits in "silence" might in fact tell a complex and revealing story about spectation, togetherness, solitude, expectation, and art. It simply depends on the subjective sense your brain makes of a situation which is inherently open to interpretation. This is precisely what I mean by delusion . . . the human mind's tremendous ability to make meaning, not just out of sensory information we perceive, but out of abstract ideas that have nothing to do with sight, smell, taste, touch, or hearing.

Several years ago, two friends of mine (Mark Lythgoe and Mark Miodownik) and I had the privilege of being the first scientists to exhibit "art works" at the Hayward Art Gallery on the South Bank in London. It was part of the Dan Flavin retrospective, an artist we touched on earlier when discussing Chevreul and the use of the perception of color in creating art. My friends and I were able to take over an entire room of the exhibition . . . in the back. My installation was a massive piece of clear Plexiglas suspended from the 20-meter-high ceiling, one meter from the ground. Embedded within this larger piece were small squares of white Plexiglas pieces that gave the impression of a floating grid of squares. This hanging piece was set in front of a wall by about 1.5 meters. On that wall was a tremendous stretched white canvas. When one looked at it from the front, it was effectively my own version of *White on White*, as the white blocks were set against the white canvas behind.

But that wasn't all. Across the room were suspended five large stage lights, all but one with different colored gel filters. The middle one was white. A computer hidden on a ledge above controlled the lighting so that it would sequence red with white, blue with white, green with white, and yellow with white. The result was that each suspended white square would cast two shadows onto the canvas. In one shadow only colored light would be hitting it. In the other shadow, only white light would be hitting it. Naturally, the colored shadow appeared the color of its illumination since that was the only light that was hitting that particular part of the canvas. What was remarkable was that the part of the canvas that only white light was hitting (i.e., the "white shadow") did not appear white at all, but instead appeared opposite in color to the other shadow. So if the white light and red light were both illuminating the canvas, then the "white" shadow would appear green! As the white squares hung in relief in front of the white canvas, this was my 3-D version of *White on White*, while simultaneously referring to Goethe's shadows to play with the notion of white itself—or, as Dan Flavin called them, the "incidental colors" that would appear in his pieces where no color was in fact present. (Note that you can recreate this on your own with two regular desk lamps, a gray gel filter covering one lamp and a colored gel filter covering the other. Point them at a white wall so that their illumination is overlapping. Then put your hand in the way of both, creating two hand-shaped shadows cast on the wall. The colors you see are a reminder that context is everything.)

The night of the opening was terribly exciting, but it turned into a bit of a disaster. The people in the Hayward were having terrible issues with the Flavin bulbs popping and circuits shorting out. My computer ended up crashing, which meant none of the lights illuminating the piece went on. Hence, no shadows;

indeed no illumination at all. But not only was the failure going to feel bad, it would also be very public. Hundreds of people were there, the "whole art world." And my piece was . . . well . . . "out of order." That is, until people viewed it—since "not working" is really a matter of opinion (it is an art gallery, after all).

There I was at the back of the black space, leaning against the back wall, despondently wondering if I should put up an Out of Order sign. To my right was a large entrance to our space (which was wholly painted black except for my white canvas), and seeping through this opening was a gentle light that cast itself onto my piece. The result was a very, very subtle shadow cast onto the white canvas. Just then two wonderfully artistic-looking men walked through that door and stopped in front of my piece. They stayed . . . and stayed . . . looking . . . moving . . . thinking. Then one started "explaining" it to the other, and the description was remarkable: "Notice the subtle play of the light, yet the shadows creating form in the relief of one against the other, suggesting a representation of difference in similarity, in the beauty of subtle difference. It's gorgeous!"

Of course the piece was, in fact, out of order, but their experience of it wasn't. Their ability to re-mean the experience re-deemed (and indeed re-meaned) the experience for me. Of course they saw a meaning in the piece. Why should they assume it was out of order? Their brains created meaning, so they didn't see a piece that had malfunctioned.

Delusion saved me.

As my experience with the Dan Flavin–Goethe–Malevich–Berkeley tribute demonstrates, delusion is one of the brain's most powerful and inescapable tools. It is the very reason why you are understanding this very sentence. As I mentioned in the introduction, my intent in writing this book is to raise an awareness in you about the source of your own perceptions, and thus create a new layer of meaning in your consciousness that will allow you to see your world and life in a new way in the future. This active mental and emotional engagement begins the process of seeing differently. Yet beyond the realm of ideas, you also are quite *literally* delusional. Not only do you not see reality— **you see things that aren't there**. This is a good thing. Just watch.

Yet beyond the realm of ideas, you also are quite literally delusional. Not only do you not see reality—you see things that aren't there. This is a good thing. Just watch.

You will have noticed by now (probably) that on every right-side page of this book there is a little diamond in the bottom right-hand corner. It's not meant as an adornment—or not *just* as an adornment. The diamonds are a flipbook.

What I want you to do is use your thumb to flip through the entire length of the book, in essence creating a small, pared-down animated movie. The diamond will at first spin to the right, because that is the direction it is drawn to spin in. After doing this once, I want you to thumb through the diamond flipbook again, only this time imagine it spinning in the other direction, to the left. It may take a few attempts, but if you blur your eyes and look around the diamond and imagine it flipping directions . . . it will.

You just saw your delusionality in action. The diamond spins back and forth depending on how you imagine it; if you imagine looking down on the central plane it will spin to the right; if you imagine you're looking up at the same central plane it will now spin to the left. *You are changing what you perceive.* To put it another way, since your brain didn't evolve to see reality, this gives you direct agency in what you *do* see.

As you just witnessed firsthand, thinking about your perception has the ability to alter it. Note that of course the motion you saw in the diamond doesn't actually exist either. So you're not only taking a series of stationary images, you're taking the small differences between them and seeing that difference as movement (a phenomenon called phi-motion). You're seeing the empirical significance of the change, not the change itself. This is a delusion without which the cinema wouldn't exist. What is more, you're also *switching* the direction of this imagined spinning depending on how you think about it. Sounds kind of like you're on drugs!

Remember: we're just talking about simple motion here. What about something more complex? The diamond flipbook is only one exceedingly simple example of our neurological tendency toward delusion. This explains the "phantom vibrations" many cell phone users report feeling, and why some hardcore video gamers have auditory hallucination-like experiences days after unplugging, known as Game Transfer Phenomena.[38] Imagine what else is possible.

Before we see how delusion feeds into seeing differently, it is absolutely crucial to understand the difference between what is happening inside the brain when you're perceiving real things versus imagined things. Why is this so important?

Because there is little difference . . . at least not qualitatively so.

Stephen M. Kosslyn is a genial, white-goateed former Harvard professor of psychology and the founding dean of the Minerva Project, a daring innovation in American higher education. His groundbreaking research with fMRIs has fundamentally changed the way scientists across disciplines conceive of perception, specifically imagined imagery versus visual imagery. His breakthrough was to prove that for the brain, *imagining things visually is no different from seeing them.*

You may be familiar with this concept if you follow sports training techniques. Many elite athletes, for example, spend serious time visualizing the sports they play, from Olympic bobsledders to professional golfers to soccer stars. This isn't a new approach, but until recently there was no hard science backing up its effectiveness. *Motor imagery* is a "mental simulation" or "cognitive rehearsal" of something you do without actually doing it.[39] Yet research shows this practice can have impacts far beyond sports.

"In psychotherapy, phobias can be treated by using mental images as stand-ins for the actual object or situation, which then leads one to become 'habituated' to the experience of actually perceiving it," says Kosslyn. "One can run 'mental simulations' and thereby practice for a future encounter. One can use such simulations to decide how best to pack objects in a trunk or arrange furniture. One can use such simulations in problem solving, to discover new ways of approaching the problem. One can visualize objects or scenes, which will improve, often dramatically, how well they are remembered. . . . Both auditory and olfactory imagery have been investigated, and in both cases considerable overlap has been found in brain areas that are activated during perception and mental imagery." In other words, without risk, you can apply mental imagery in every realm of your life—to help overcome social anxiety, dispatch a debilitating workload, or win at your weekly poker night. Likewise, researchers are bringing mental imagery to bear in diverse areas, such as doctors who are incorporating it into stroke rehabilitation.[40]

For the brain, then, "realness" is much more capacious and open-armed than our more narrow and conventional conception of *real* as physically experienced and *not-real* as imagined. On a neurocellular level, they *are* both physically experienced. With this "in mind," the effects of internal imagery (and delusion) are really quite intuitive. Think about it in terms of sexual arousal. Fans of *The Big Lebowski* may recall when the porn king Jackie

Treehorn says, "People forget that the brain is the biggest erogenous zone." What he means is that imagining a naked body can create as much libidinous excitement as that same body right in front of us. Real sex and imagined sex (and, increasingly, virtual reality sex) create blood flow in the same parts of our brains—and of course in other areas of the body as well.

So how does mental imagery play into creative perceptions? The answer comes down once again to the history of usefulness our brain encodes, and how this perceptual record determines our future seeing. The fundamental truth about perception that I explained hasn't changed: We don't see reality—we only see what was useful to see in the past. But the implication of the brain's delusional nature is this: The past that determines how you see isn't just constituted by your lived perceptions but by your imagined ones as well. As such, you can influence what you see in the future just by thinking. The link between the two is that what we see now represents the history of what we saw before, imagined or otherwise (though not all having the same weight).

This is why we are the experiencers *and* creators of our own perceptions!

Think back to Ben Underwood and his astonishing ability to echolocate. He conducted an intense trial-and-error process to adapt and learn how to "see" through his clicking, and this changed the structure of his brain. A similar process happens with mental imagery, only the trial and error takes place inside us instead of outside. Strengthening the "muscle" of your brain means seeking out not just enriched physical environments, but mentally constructed ones too. Imagined perceptions shape your future much like lived ones, as both physically change your neural architecture (though to differing degrees, since lived experiences generally have a stronger statistical impact), and not always for the positive. Consider rumination where one becomes "trapped" in cycles of reliving the negative meaning of an experience (but not the experience itself), thereby strengthening the wiring accordingly, making its meaning disproportionately significant.

If each time we do one of these embodied, low-risk thought experiments they are being encoded as if they were experiences, then we need to reconsider the big takeaway from the last chapter: Change your ecology and you change your brain.

Change your ecology and you change your brain.

Imagined perceptions are so influential for seeing that they are a part of our environment. Which is another way of saying that context is still everything, *but an absolutely essential part of that context resides in you.* You are your own context. Your brain matches the complexity not just of its external surroundings, but of its internal environment as well. If you imagine complex, challenging possibilities, your brain will adapt to them. Much like a tiger trapped in a zoo exhibits repetitive displacement behavior, if you cage your imagination within the bars of the dull and neurotic, which often portray one's fears more than they do an empirical "truth," then your brain will adapt to these imagined meanings, too. Like the sad tiger pacing back and forth within its puny cage, your brain too will ruminate cyclically upon destructive meanings, and in doing so make them more significant than they might need to be. This present perceptual meaning becomes part of your future history of meanings, together with the meaning (and re-meanings) of past events, thus shaping your future perception. If you don't want to let received contexts limit possibility, then you need to walk in the darkest forest of all—the one in your own skull—and face down the fear of ideas that challenge.

You must learn to choose your delusions. If you don't, they will choose you (remembering, of course, that not all delusions are choosable).

Your brain is, at its core, a statistical distribution. Thus, your history of experiences creates a database of useful past perceptions. New information is constantly flowing in, and your brain is constantly integrating it into this statistical distribution that creates your next perception (so in this sense "reality" is just the product of your brain's ever-evolving database of consequence). As such, your perception is subject to a statistical phenomenon known in probability theory as *kurtosis*. Kurtosis in essence means that things tend to become increasingly steep in their distribution . . . that is, skewed in one direction. This applies to ways of seeing everything from current events to ourselves as we lean "skewedly" toward one interpretation, positive or negative. Things that are highly kurtotic, or skewed, are hard to shift away from. This is another way of saying that seeing differently isn't just conceptually difficult—it's *statistically* difficult. We're really talking about math when we say, "The optimist sees the glass as half full and the pessimist as half empty," though in my view maybe **true optimists are just glad to have a drink in the first place!**

The good news is that kurtosis can work to your advantage

> *You must learn to choose your delusions. If you don't, they will choose you (remembering, of course, that not all delusions are choosable).*

(although this depends on your choosing a useful delusion, since you're just as capable of choosing one that is destructive). The linchpin is to exploit your ability to influence your inner probability distributions. Imagined perception is self-reinforcing, so by taking control of it you can alter the subjective reality that your brain's interpretations create. Think positively today and it's statistically more likely you will do the same tomorrow.[41] This seemingly "soft" truism has in fact been borne out by hard science. The psychologist Dr. Richard Wiseman has pioneered the systematic study of chance and luck, and has found that people who are consistently "lucky" in life share patterns of believing things will work out well in the end; being open to experience; not ruminating on disappointing outcomes; taking a relaxed approach to life; and looking at mistakes as learning opportunities.[42] It will likely come as no surprise that the same self-reinforcing perceptual and behavioral momentum also applies to negativity.

It is worth noting that in spite of the empowering possibilities of imagery and the imagination applied in the present, or in re-meaning the past, they do have constraints. For example, while seeing the color blue is itself a meaning, you can't imagine blue to be a different color than it is. That is one perception you can't re-create, since it is hardwired into your brain from evolution. So, clearly, our consciousness operates within bounds. It's just that those bounds are often set artificially, usually by others, but even more troubling when set by ourselves. We've all had the experience (kids more than anyone) of asking someone a question and the answer is no. Hence the question . . . *Why?* The question seems perfectly reasonable, and yet, when challenged, the person answers, "Because that's the way it is!" or some other clichéd, ill-thought-out answer. None of which have anything to do with the physics of the world, and instead have everything to do with the assumptions encoded in that person's brain. I call this the "Physics of No," since the answer is treated as if it's a law of nature. How we learn to overturn the Physics of No will come later in this book. It is not easy, and it takes courage to imagine, think, and challenge not just other people, but ourselves.

What's fascinating is that we are predisposed to different directionalities of perception . . . different delusions . . . at different ages. Recent research shows that adolescents tend to look at the world so that their seeing confirms their emotional state. If they are sad or anguished (as many of us will recall from personal experience is common at this age), they will seek out sad images, and interpret their environment with a sad/anguished bias. In academic science-speak, adolescents have a harder time mastering "emotional regulation."[43] This is the neurological explanation for what we often jokingly call teen angst, and without it we likely wouldn't have ripped jeans, the term "emo," or bands like Nirvana. While such angst is in many ways a rite of passage, both culturally and personally, it is critical for young people with a high delusion toward negative statistical distributions to learn how to develop "regulatory skills," as suicide is the third-leading cause of death for males and females aged 10 to 24 in the United States.

The good news, however, is that as we get older we tend to look at things that represent the emotions we want, which is a kind of aspirational probability for our brain to sculpt future perceptions with. This is why, on a bad workday, we might find ourselves googling "dream vacation" (and not "closest bridge to jump off"). Yet in spite of our behavior tending to get "wiser" with age, this doesn't mean we should exert less agency in choosing our delusions. Circumstantial conditions (the death of a loved one, a divorce, unemployment, a midlife crisis) can make us vulnerable to useless thinking in the present, which of course leads to more useless perceptions in the future. Plus, each one of us has our own distinct statistical "constitution" in our brain. Some people are more prone to be healthy regulators; others turn out to be non-regulators. But as you'll recall from my story of the dinner party experiment in the introduction, we can be easily primed (or even prime ourselves) into higher (or lower) states that affect our behavior.

A concept crucially connected to this facet of perception is *confirmation bias*, also known as "myside bias," which likely gives you a hint what it means. It's really a rather self-explanatory idea, if not very flattering about the human propensity to listen to no one but oneself, much less about our collective deafness to nature itself. Confirmation bias describes the tendency to perceive in such a way as to confirm your already established point of view. It is present in everything from the way you argue (and what you take away from arguments) to how you behave in relationships and at work. It even structures what you remember, shaping your memories according to your ideas about yourself (which are often incorrect). The concept need not be restricted to individuals, but also applies to social groupings. Political parties, jingoism, sports fandom, and religion all suffer from *cognitive bias*. Such biases also shape the history of whole peoples, as is the case with the sexes. Historically, the prominence of women (educationally, professionally, civically) in Western society was retarded in several ways not just by the mis/disinformation that men promulgated about their abilities, but also by women's own internalization of it.[44]

Confirmation bias takes us back to the fundamental premise of this book: We don't have access to reality . . . we perceive (or indeed, in this case *find*) the version of reality with which we are familiar . . . and often that makes us look good. For instance, if you ask a large room of people (especially men) who are a random sampling of the population at large, "How many here are better than average drivers?" the vast majority typically raise their hands. Sorry, this isn't possible. Not everyone can be above average. By definition, assuming a uniform sampling, half of the population has to be less than the median. That's the essence of a median. Likewise, studies show that we believe ourselves much more willing to engage in selfless, noble acts than we actually are (while at the same time underestimating the generosity of our peers).[45] *We are our own best heroes!*

Yet our blindness to our biases makes it exceedingly difficult to notice them. In a now-famous study published in 2012, Hajo Adam and Adam Galinsky found that people who wore a white doctor's lab coat while executing a series of mental exercises that tested attention did better than participants who wore street clothes. They also did better than participants who wore the same lab coat but were told it was a painter's coat rather than a doctor's.[46] This area of research is called "enclothed cognition" and shows that not only do others project expectations onto us according to what we wear, but we ourselves project similar expectations that directly influence our perception and behavior—yet another instance of delusion. Adam and Galinsky's experiment is a powerful demonstration of the effects of *priming*: one stimulus (wearing the lab coat) influencing behavior and perception when confronted with a subsequent stimulus (the attention test).[47] So not only do we "brand" ourselves for others and condition how they treat us by what we wear each day, we are also engaging in potent *self*-branding.

It probably won't surprise you to learn that priming, or subliminal stimulation, is of great interest not just to scientists but to marketing researchers as well. While it's frightening to think we're not the masters of our own buying impulses (let alone a thousand other behaviors), it has been shown that people are more likely to buy French wine while shopping if French music is playing.[48] Outside of artificial settings, in the "wild" of daily life, the same thing happens. Coins seem bigger to poor children than to children from better-off backgrounds. People see hills as steeper when already physically tired. Distances appear longer if one is carrying a heavy load. We generally perceive desirable objects as closer than they actually are. Obviously, then, something powerful is happening in the brain on very basic levels of perception—which is why The Lab of Misfits created an experiment to explore just this at the London Science Museum in 2012, during our club-like open nights for the public we called "Lates."

The experiment was run by one of the misfits in the lab, Richard C. Clarke, and in fact you're already partially familiar with it. Remember the brightness-contrast "illusion" from Chapter 1, in which you perceived the same shade of gray differently depending on the color-context that surrounded it? We used the same "test," only our aim wasn't to demonstrate that the participants didn't see reality. Instead, we wanted to study how a sense of power affects perception. We already knew that delusion strongly influences the complex tasks the brain carries out, such as calibrating one's self-esteem or deciding what to buy, but what about more basic low-level perceptions unrelated to a goal? Could delusion physiologically affect not just the seemingly more highly cognitive process of frog-princes but the more basic perceptual process of frogs as well? If so, this would mean that delusion played a pivotal role at every operative level of perception.

. Perceiving brightness is one of the most basic visual tasks the brain performs. As such, we believed it would be the right tool with which to examine our hypothesis that one's state of control or power would indeed determine low-level perceptions. The first task at hand was to find subjects to volunteer for the experiment, which meant pulling them away from the variety of scientific exhibits on display, as well as the live music and the cocktail "Brain Bar." When the recruitment was all said and done, we had fifty-four volunteers. We randomly assigned them to three different groups: the low-power group, the high-power group, and the control group. Why? So that we could prime them.

We used a common priming technique involving writing and memory. We told the participants that they were going to take part in an experiment to investigate their perception of past events. We then asked them to compose a short narrative essay. The high-power group had to write about a memory in which they had been stressed but had felt in control. The low-power group also had to write about a memory where they too had felt stressed, but one in which they had had very little control. Participants in the neutral control group wrote about how they had ended up at the Science Museum that day. Three very different primes, potentially setting up three very different delusions.

Now that the three groups had three distinct "self-biases" . . . or lack of a bias, in the case of the control group . . . we began the brightness-contrast tests (which the participants thought was unrelated to the supposed memory experiment we had led them to believe they were a part of). Each subject was shown eight different gray "targets" embedded in different surrounds. The sequence was random and the surround sets varied in color, relative location, and shape, while the spectrometer-measured brightness was constant. The subjects rated the brightness of each target numerically and in this way allowed us to quantify their subjective, priming-influenced perception of the "illusory" shades of gray their brains constructed. We then sent the participants back out into our misfit "Night at the Museum" lab, as we quietly speculated about what we might learn. Sure enough, after analyzing the numbers, we had an extremely revealing finding.

While previous research has shown that individuals in low states of power demonstrate reduced abilities with complex perceptual tasks, we discovered that the opposite was true for a basic process like perceiving brightness. On average, the subjects primed into the low-power emotional states used the contextual visual cues to a greater degree than those in the high-power and control groups. The ability of the low-power group's brains to make sense of the senseless stimuli was *sharpened* by the delusion we had planted that made them feel less in control. Their evolutionary need to gain back a sense of power (i.e., control) and thus construct a useful perception kicked in more strongly than it did for the other groups, allowing them to see differently than their peers. Which means they behaved much like children did, who also saw the illusion more strongly. Children were, in a sense, more willing to "believe." As such, their perceptions and realities were significantly, systematically, and statistically altered by their psychological state. In short, people in a lower state of power generate behaviors that are an attempt to increase their power.

Power even alters the ways you look at a scene by altering your eye movements. People in different states actually look at things differently. For example, people with lower power look at background, whereas those with higher power look at foreground. Since what we look at determines the statistics of the images we sense, power alters the orientation of our windows, as well as the way we move our mobile homes around the world. While it may seem slightly frightening to realize how suggestible we are, and how the delusional nature of perception can make us either "lucky" or "unlucky," these behaviors make sense. If one is in fact in a low state of power, then one's behavior should orient toward regaining control. Thus, research shows that even "background mood states" strongly affect our decision making.[49] The point is that our brain, while amazing in so many ways, also makes us vulnerable, but that vulnerability can itself have utility. The coming chapters are about how to actively use your brain to help and not hamper you—to give you new ideas and unforeseen creative insights. We are able to do this because we have consciousness—a brain that modifies itself.

To break down your own personal Physics of NO that the meanings applied to your past experiences have given you, you must embrace one fundamental fact about yourself: You have assumptions! They are deep assumptions, so deep that you can't move without them—not even one step forward. You die without them. These assumptions are a result of your empirical history of trial-and-error experience, whether that experience be internal or external. They necessarily limit what you think, do, and feel. But in doing so they can destroy relationships and careers. Deviation from self-sabotaging or even just familiar behaviors begins (but doesn't end) with a disarmingly simple first step: awareness. Awareness that you have assumptions, and that you are defined by them. And I don't mean, "Yeah, I know we *all* have assumptions . . . but . . ." I mean really knowing this, which means actively *living the knowing*. Having this kind of awareness, as we'll see, is a revolutionary tool. It is where you start changing not just your brain and your perceptions, but your whole way of being as a human.

CHAPTER 6

The Physiology of Assumptions

*Can you think
of a colorless
apple?*

Delusion is an important tool for creating new and powerful perceptions because it allows us to change our brains—and thus future perceptions—from the inside. But if the human brain is a physical representation of the history of trial and error from evolution to learning and all perceptions are reflexive responses, then how can anyone, even those of us with the most grandiose delusions, change our perception? After all, the past, as we all very well know, is stubbornly unchangeable. What happened, happened. Yet it is not that simple when it comes to the workings of the mind, since, as we also know, we're never recording reality, much less the reality of time itself.

What our brains carry forward with us into the future isn't the actual past . . . definitely not an objective one. What your perceptual history of reality gives your brain are **reflexive assumptions** manifest in the functional architecture of the brain with which you perceive the here and now. These assumptions determine what we think and do, and help us to predict what to do next. It is important to note also the opposite: they also determine what we *don't* think and do. Independent of context, our assumptions are neither good nor bad. They are simply us . . . collectively and individually.

We are very lucky that our brain evolved to have assumptions, yet many often seem like the air we breathe: invisible. When you sit down, you assume the chair—usually—won't give way. Every time you take a step, you assume the ground won't give way, that your leg won't give way, that your leg is far enough in front of you, and that you've changed the distribution of your weight sufficiently to propel yourself forward (since, after all, to walk is in fact a continual process of falling). These are essential assumptions.

Imagine if you had to think about walking, or breathing, or all the other eminently useful yet thoughtless behaviors your brain lets you effortlessly perform. Likely you would never move. This is partly because attention can only be directed at one task (what we call "local" information in perceptual neuroscience), but also because of priorities: If you had to think about every single operation underpinning your existence, you would probably want to spend most of your time thinking about keeping your heart beating and your lungs breathing, making sleep effectively impossible. The fact that you don't have to consciously keep your heart beating is thanks to your brain's role as the command center managing built-in physiological assumptions for your body. Having to expend significant thinking energy on such a task wouldn't have been advantageous for survival in a shifting world. As such, we didn't evolve to perceive in this way.

So what actually guides our perceptions . . . what is it about the past that is being extracted? An answer: A set of baseline mechanical assumptions that our species developed over many, many millennia to right this very moment. This goes not just for breathing, but for sight as well. We—like other animals—are born with many assumptions (such as physical laws) already "bred" into us. This is why our eyes can't suddenly be reprogrammed to have the vision of a stomatopod; we developed only to process light in the way that worked best for our species. Yet not all of the preset assumptions in the brain are so basic (basic, that is, in the sense of basic functions; they are obviously very complex). This is because we are not only like frogs. We are also like turkeys.

Turkeys come into the world with an endogenous reflex to protect themselves when their retinas receive a pattern that is consistent with birds of prey, even without previous visual experience. In a curious experiment from the 1950s measuring young turkeys' fear responses, a silhouette of a bird of prey scared them while one of a duck did not. They just knew.[50] Likewise, recent research suggests that humans are born with an innate fear of snakes, an adaptive assumption from our past that helped us survive, and still does. We inherited this from ancestors we'll never meet. Likewise, in a study conducted at the University of Virginia, researchers tested the speed with which preschool children and adults reacted to different visual stimuli. Both the children and the adults showed an "attentional bias" toward snakes, detecting them more quickly than they did unthreatening stimuli such as frogs, flowers, and caterpillars.[51] It is clear, then, that human beings are not a blank slate.

The concept of the blank slate, or *tabula rasa*, forms part of an age-old debate over how humans become who they are and end up with the lives they lead. Everyone from philosophers to scientists to politicians has argued over the topic, since it has implications about essential ethical questions that underpin how best to create equality in a society. Are we products of nurture or nature? Do we come into the world with our personalities and physical constitutions already in place, or do our experiences and circumstances shape us? If we knew the answer to this question, the thinking goes, we could better address the ills in our societies. But it's the wrong question, as the field of neural development has shown, in particular the field of epigenetics: It's not one or the other. Nor is it a combination of both nurture and nature. Rather, it's their *constant interaction*. Genes don't encode specific traits as such; rather, they encode mechanisms, processes, and elements of *interactions* between cells and their cellular and non-cellular environment. Genetics and development are inherently ecological processes.

This insight, which has gained currency in neurogenetics, is apparent when you study development inside the brain. What is inherent in the brain are ways of growing and a rough blueprint of what is to be grown. But the exact type of growth that occurs is surprisingly malleable. If you transplant a piece of visual cortex into the auditory cortex, the transplanted cells will come to behave as if they are auditory cortical cells, including establishing connections to other auditory areas, and vice versa for the visual cortex. For instance, the transplanted visual cortex will establish connections with a different nucleus within the thalamus than it would have if it remained in the (primary) visual cortex. It will also form connections with different cortical areas than it would have if left in the visual cortex. Even its internal process structure changes. For instance, whereas cells in the middle layer of the visual cortex will become dominated by connections from *either* the right eye or the left eye (called ocular dominance columns), this pattern of connectivity does not form if a region of naïve visual cortex is transplanted into the auditory cortex. It is a combination of the properties of the cells, the ensemble of cells, and their connections that determines the function of that cell and its larger role in the "community" of cells (which is much like a social network). This physiological reality is also a biological principle: Systems are defined by the *interaction* between their inherent properties and their external relationships in space and time . . . whether cells in the cortex, or a person in a larger society or organization. This means that the "meaning" of each of us is necessarily defined by our interactions internally and externally. Thus, like us, a developing visual neuron is largely *pluripotent* (i.e., capable of different potential uses) within the realm of a particular cell time (much like personality traits). Like us, a neuron becomes defined by its ecology. Yet this contextual elasticity doesn't mean that our slate is blank. On each one of our tabulae is written the same bedrock assumption that to perceive and survive we must have assumptions.

What is more, also encoded is the assumption that we will look for *more and more* assumptions.

Your brain uses experience to acquire as many of these assumptions as possible in the hopes of finding principles that can be applied across contexts (much like a theorem in physics). For example, heights: Oddly enough, it appears that we are not born with a fear of heights and an awareness of the dangers they hold. Recent research using "visual cliffs" like those used to test the Kittens in a Basket has shown that while infants avoid heights, they don't exhibit an autonomic fear response.[52] However, as we develop over time we learn to respect heights, thanks to an added embedded hierarchy of assumptions we gain from lived experiences, whether from the time our parents yelled at us for going near a cliff or the time we fell off a top bunk and hurt ourselves. Whatever the origin of our caution, henceforward we carry a very useful assumption that keeps us safer. It feels like common sense, because it is, but it wasn't in our brain from the start. Other low-level assumptions—*thousands* of them, actually—that influence our behaviors relate not to physical survival but to social survival, yet they too are very physical.

The mechanics of your eye movements should be the same as those of every other human on the planet, right? We each have the same visual processing hardware in our brain, after all, so we should all use the same software. It's an intuitive conclusion, but it's wrong. We use different "programs" to implement our seeing, depending on where we're from. In a delightfully revealing experiment from 2010, David J. Kelly and Roberto Caldara discovered that people from Western societies exhibit different eye movements than people from Eastern societies. As they put it, "Culture affects the way people move their eyes to extract information in their visual world." Asians extracted visual information more "holistically," whereas westerners did so more "analytically" (there was no difference in their ability to recognize faces). Western cultures focus on discrete elements or "salient objects"

with which they can confidently process information, as befits a markedly individualistic culture. Eastern cultures, on the other hand, put higher value on groups and collective goals, causing them to be drawn to a "region" rather than one specific feature of a face. In practice, this means that Asians focus more on the nose area on average, while westerners are drawn to eyes and mouths. Different eye movements shape perception dramatically, as what we "look at" constrains the nature of the information that the brain then uses to make meaning. Changing the nature of the input differentially constrains the potential meanings. Socially learned assumptions and biases like these affect our gray matter, as well as subsequent perceptions and behaviors, yet they are developed so unconsciously from larger cultural assumptions that we don't even know they are there, embodied in our brains.

Other very important assumptions that shape our perceptions, and even our life trajectories, are also learned socially, but their effects are more easily detectable in our behaviors than subtle eye movements. This is most powerfully illustrated in that context from which every human originates, and which we all share in some shape or form: *family*.

Meet Charles. He was the fifth child born into a family of six children in Shropshire, England, in 1809. He was a cute British lad from a well-off home and had rosy cheeks and straight brown hair. He liked being outside among trees and nature. He collected beetles and had a precocious knack for natural history. Unlike some of his older siblings, he was prone to questioning conventional wisdom; for example, if it truly mattered that he do well in school, which irked his father. Charles loved asking questions just for the sake of asking them, as crazy as they might sound or in fact be. In 1831, landing on an unconventional path that also bothered his father, he set off for South America on a ship called the *Beagle*, in the hopes of making valuable observations about geology, insects, marine life, and other animals. His observations, indeed, were valuable. Through his study of twelve finches in the Galapagos Islands, and by questioning his own beliefs about supposed divine design and the origin of our species, he changed science forever. Charles, of course, was Charles Darwin, the man who gave us the theory of evolution.

While Darwin was indisputably brilliant—not to mention fanatically curious—he might not have uncovered the evolution-ary process if it hadn't been for the assumptions that being one of the younger siblings in his family, or a "laterborn," gave him. This is one of the arguments of the evolutionary psychologist and MacArthur "Genius" Frank Sulloway. In his research, on which he continues to build with powerful new studies, Sulloway has shown that where you fall in the birth order of your family strongly affects your personality traits, behaviors, and percep-tions. This is because we are all competing to win the time and attention of our parents, and depending on where you fall in the "sibship," you develop different strategies and biases for succeed-ing at this. As Sulloway memorably describes it, "The result has been an evolutionary arms race played out within the family."[53] This doesn't mean that siblings are consciously fighting their older and younger siblings in a cutthroat Darwinian battle for mommy and daddy's love; rather that the structures of families inevitably affect who we become, since we get better at different things as a function of behaviors that more naturally fit our birth order. Firstborns, for example, tend to win their parents' favor by taking care of their younger siblings to some degree, and thus generally develop a high degree of conscientiousness and respect for authority figures. Laterborns, in contrast, strengthen "latent talents" that will garner them parental attention. For this reason they're more open and adventurous, and less reverent of author-ity. How each sibling must act in order to "succeed" inside the ecology of the family becomes a matter of perceptual assumptions, recorded in the brain through a history of trial and error that has its own kurtotic statistical weight.

So, is the point that Charles Darwin saw differently simply because he was Darwin kid number 5? (Interestingly, Alfred Russel Wallace, a scientist who was developing ideas about evolution similar to Darwin's at around the same time but who published less, was also a laterborn.) Or is it that if you're a first-born or only child, you should just call it a day and forget about being innovative? The answer is no to both. The point is that the assumptions your brain has developed from multiple sources of experience over different timelines don't just shape your perception . . . they *are* you. They are the layers of "empirical significance" that you have attributed to stimuli, which define the reality you perceive . . . how you perceive yourself and others, and thus how you lead your life. But what are assumptions, physically speaking, in the brain?

We know now that all perception ultimately comes down to deciding whether to go toward or away from something, and this even crosses species. This going-toward-or-away is an essential part of why we perceive what we do, and assumptions inescapably influence the direction we choose to go in. So how does this process create our perception?

Assumptions are deeply physiological … electrical, in fact. They are not just abstract ideas or concepts. They are physical things in your brain, with their own sort of physical "laws." This is what one could call the *neuroscience of bias*. The reality we see projected on the "screen" of perception begins with the flow of information our five senses take in. This stimulus (or stimuli if there are more than one) creates a series of impulses at your receptors that move into your brain (the input), becoming distributed across the different parts of your cortex and other areas of your brain until they eventually settle on activating a response (motor and/or perceptual . . . though the separation between motor and perceptual isn't as distinct as was once thought). That is basically the whole of neuroscience in one sentence, with an emphasis on "basically." Perception **is** nothing more than a complex *reflex arc*, not unlike that which causes your leg to kick out when the doctor hits your patellar tendon underneath your knee-cap. At bottom, our lives are in fact nothing more than millions and millions of sequential knee-jerk reflexive responses.

What you experience at any moment is just a stable pattern of electrical activity distributed throughout your brain—an unromantic view of perception, but it is nevertheless roughly accurate. Over the course of your life, the electrical patterns generated in your brain in response to stimuli become ever more "stable," which is called an *attractor state* in physics.[54] Dunes in the desert are an example of an attractor state, as are whirlpools in a river. Even our galaxy is an attractor state. They all represent emergent, stable patterns that arise from the interactions of many individual elements over time. In this sense, they have their own stable energy state, or *momentum* (in that they can be difficult to shift), that is most natural to continue in (though the brain states of children are not as stable as those of adults). What evolution does is to select some attractor states or, more accurately, a sequence of attractor states—as more useful than others.

The electrical patterns are created by the neural pathways that link the different parts of the brain . . . a staggeringly complex, sprawling superhighway of connectivity that is your brain's infrastructure. These patterns make some behaviors and thoughts very probable, and others not. Studies have shown that the more connectivity you have in this superhighway, the more likely you are to have more diverse and complex assumptions (e.g., stronger vocabulary and memory skills).[55] Yet in spite of the brain's

prolific interconnections and the importance of these interconnections in your perceptions, the number of neuroelectrical
pulses that your perception actually entertains and uses throughout its life is very small. That is, *relatively* speaking, because their
potential is nearly infinite.

The cells in your brain constitute the **Cartesian You**. By "Cartesian," I'm referring to the French philosopher René Descartes,
who espoused a mechanistic view of human consciousness, out
of which came his famous phrase *cogito, ergo sum*: "I think, therefore I am." Your thinking, and therefore your being, depends on
the cells that constitute your brain's railroad system, allowing
the electrical patterns—like trains—to follow their reflex arcs.
Counting the number of these cells is itself an interesting story.
For years, neuroscientists have cited and re-cited that there are 100
billion neurons in the brain. (Neurons are nerve cells that receive
and send signals through synapses in the nervous system.) It's a
nice, round, weighty number. It also happens to be wrong.

No one appears to know where the figure of 100 billion first
emerged, and each scientist who cited it seems to have assumed
it to be correct for a terrible if fairly understandable reason:
because they'd heard it from someone else. With wonderful
irony, it more likely reflects an inherent bias we have to whole-
ness, and so in this case the round number of 100. This changed
in 2009 when the Brazilian researcher Dr. Suzana Herculano-
Houzel implemented a remarkably clever innovation and proved
that the figure was an erroneous assumption . . . a received
idea that had inadvertently disguised itself as a fact, a scientific
meme.[56] (More on memes shortly.) Through a brilliant research
method that involved liquefying four brains donated to science,
Herculano-Houzel discovered that on average we have 14 billion
fewer cells than we thought, which is itself the number of cells
in a baboon's brain. Though not an insignificant reduction, the

86 billion we do have is still a *lot* of neurons. So the thinking-therefore-being you *is* all these neurons and how they talk to each other (and of course with the rest of your body and environment, lest you think that you are only your brain).

Now let's return to the relatively tiny number of electrochemical patterns your brain engages when you perceive. The cells that make up your brain form 100 trillion connections. That is a massive number, but what does it really mean? Since the potential connections form the possible reflex arcs that shape behavior, what's really at stake is how you will respond . . . what you perceive and whether the perception you produce will be good or bad, innovative or complacent, risk-taking or conservative. So what we're talking about is *possible* responses versus *actual* ones, and the possible ones are almost inconceivably numerous. For instance, imagine humans had only 50 brain cells instead of 86 billion. (Ants have approximately 250,000 brain cells, so having 50 would make you an exceedingly elementary organism.) Yet if you took those 50 brain cells, each with 50 connections, and came up with all the possible ways that they could connect to each other, the number of different possible *connectomes* (as a set of connections is called) would be greater than the number of atoms in the known universe. Just 50 neurons! Now consider all the potential patterns that 86 billion cells forming 100 trillion different connections could make. The number is pretty much infinite. Yet our perceptions in practice *aren't* infinite—far from it, in fact. They are a mere minuscule subset of what is objectively possible. Why? Because of our assumptions, which come from experience.

These experience-driven biases define and limit the synaptic pathways through which our thoughts and behaviors come into being. Hence, the relationship between a stimulus (input) and its resulting neural pattern (output) that *is* a perception is constrained by the network architecture of your brain. This electrochemical structure is a direct representation of the empirical process of shaping the brain by trial and error. It is a grid of possible responses shaped by experience, from seconds to millennia ago, from enriched environments to impoverished ones. Our reflex arcs, then, are contained not just in our bodies as such, but in our ecology. Layers upon layers of history passed down into you, which means that most of the experiences that shaped your "frog" (and turkey) brain happened *without you even being there.* Yet this evolutionary history is what determines the "reality" of so much of what you perceive and how you behave. Combine this species-level experience with your own lived empirical history, and you have your own unique tapestry (or more accurately, embedded hierarchy) of assumptions that allow you to survive at the same time that they threaten to limit the electrical currents— that is, the ideas—that produce your responses.

In short, your assumptions make you you. This means pretty much everything you perceive about your own conscious identity would be on the line if they were ever to be doubted. Yet the *process* of creating these brain-based biases that give you your "you-ness" also endows us with the unique people the world so badly needs (whom I'll refer to as "deviators").

Near the end of 2013, an infant boy in Guinea became infected with the Ebola virus, an extremely painful and contagious hemorrhagic fever with a fatality rate of approximately 50 percent. After this "index case"—the first person infected—Ebola spread throughout West Africa with alarming speed and by early 2014 hit epidemic proportions, the first time this had happened since the virus's discovery in the 1970s. It spread to a total of nine countries, the worst hit of all being Liberia, where nearly five thousand people died. It was also in Liberia, in the summer of 2014, that a 40-year-old Liberian-American citizen named Patrick Sawyer visited his sister, who had contracted Ebola and died while he was there caring for her. After his sister's funeral, on July 20, Sawyer flew to Nigeria, where Ebola had yet to spread from its neighboring countries. When he arrived in Lagos, Nigeria's most crowded city, Sawyer collapsed at the airport, racked by vomiting and diarrhea. Doctors of the public health system happened to be on strike at the time, causing him to end up at the private hospital where the Nigerian doctor Ameyo Adadevoh worked.

Dr. Adadevoh was a woman who seemed to have medicine "in her blood." Her father was a highly respected pathologist and university official, and she had inherited his same rigor when it came to her profession. She had wavy black hair and large dark eyes that projected a serious gaze in the halls of First Consultant Hospital in Lagos, where Dr. Adadevoh was a senior endocrinologist. When Sawyer arrived in her hospital, she was the attending physician. He claimed that he only had a case of malaria, but Adadevoh wasn't convinced. He tried to leave the hospital, but she refused his request, convinced that he should be tested for Ebola, even though she'd never treated anyone infected with the disease before. Sawyer became very upset, and she and her colleagues ended up having to physically restrain him. During the ensuing chaos of this altercation, his IV popped out of his arm, spraying blood on Adadevoh. They were finally able to subdue him, but the struggle didn't end there.

While Adadevoh and her fellow doctors were waiting for the results of Sawyer's test, the Liberian government put diplomatic pressure on Nigeria to release him to them. Again, Dr. Adadevoh refused. With no secure transportation method available in the country to guarantee that he wouldn't risk infecting anyone else, she and her colleagues fought to maintain his quarantine—and won. By neutralizing their "patient zero," they could now focus their energies on mobilizing Nigeria's response to contain the impact of Sawyer, who had passed his virus on to twenty people.

On October 20, 2014, three months after Sawyer arrived in Nigeria and Adadevoh became his doctor, the World Heath Organization officially announced that Nigeria was Ebola-free, a spectacular and heartening feat amid so much suffering in the region. As the *Telegraph*'s chief foreign correspondent wrote at the time, "It may not sound like tremendous news, given that the death toll is still rocketing elsewhere in West Africa. But given how bad it could have been, it's something to be extremely thankful for. I could just as easily be writing now that Ebola had claimed its 10,000[th] victim in Nigeria, and was on its way to kill hundreds of thousands more."[57] As global media soon broadcast, this victory was largely thanks to Dr. Adadevoh, who tragically wasn't able to see the impact of her work. Sawyer died of Ebola, along with seven people who caught the virus from him—one of the victims being Adadevoh herself, who died on August 19, 2014. Since her death, Adadevoh has been lionized as the hero that she was—the person who, along with her colleagues, protected Nigeria from what would have surely been an extremely fatal outbreak if Sawyer hadn't been kept in quarantine. Their brave response—in the face of the assumptions of others—has become a model for other countries around the world, and an example of the right assumptions working. Adadevoh's story offers a vivid illustration of how assumptions produce useful thoughts and behaviors, and also of how these assumptions can be specific to a person creating ideas that others are unable to see.

Let's take a look at how Dr. Adadevoh's assumptions shaped her thoughts and perceptions (leaving mostly aside the process by which her assumptions came into being), to better understand how your neural patterns shape yours. To do so, it's time to introduce an important new concept central to getting more and better ideas from those 86 billion electrically charged cells in your head.

Your Space of Possibility

The space of possibility is the patterns of neural activity that are possible given the structure of your network (or connectomes). Collectively, your neural network determines all the different possible patterns inside your head. In other words, it contains the matrix of perceptions/ideas/behaviors that you are capable of having, which means it contains a lot (but again, its potential is far from infinite). These perceptions/ideas/behaviors run the gamut from the most luminous, earth-shaking ones, to the most banal, humdrum ones . . . most of which you'll never actually experience, but are theoretically possible given a certain concatenation (network linking) of stimuli. Moreover, *outside* this space are those perceptions and ideas that don't, indeed *can't*, arise in your brain and can't occur to you, at least at that moment—which arguably are *much more numerous* than the ones that are possible. Your assumptions (i.e., the connectivity between brain cells that is your history) determine the boundaries and everything that falls within—and hence the structure and dimensionality of your space of possibility. Note also that each of these patterns is related to others, some being more similar to each other and others less so.

Now, while an infinite number of neural patterns are theoretically possible, not all of them are useful. Let's consider Adadevoh's actions inside the schema of the space of possibility. She had an assumption that she shared with Nigerian government officials: that the spread of Ebola needed to be contained as quickly and completely as possible. Where she differed from others was in how to effect this, which led to her second assumption: that Sawyer needed to be kept in quarantine in the Lagos hospital, as opposed to the officials who assumed that the best response was to get Sawyer out of the country and back to Liberia as fast as possible. Because Adadevoh had different assumptions, she had a different space of possibility, which meant she had different potential states of brain activations . . . manifested as ideas, thoughts, beliefs, actions, and so on. These different assumptions encoded in the physiology of her brain represented her own unique history, which allowed her neurons to generate a 'next possible' perception (an attractor state) that people without the same wiring (assumptions) couldn't generate. Again, we are focusing here not on what that history was as such, but simply on the fact that because of it, what was possible for her to perceive was not possible for others to perceive. She was not making a "big jump" . . . for her. The "big jump" she was making was from their perspective since they were blind to what she could see, simply because they had different spaces of possibility. To begin with, Adadevoh was a seasoned professional who not only knew but *followed* medical best practices, even in a moment of conflict and danger. So she had sound convictions, but even more important, she had the courage to pursue them. In spite of pressure from the government, she put the collective well-being of her country-men ahead of possible damage to her career. As such, the value of sacrifice for the good of people beyond herself was an essential assumption that guided her perception and her actions. It shaped her space of possibility as much as her technical medical training did. In other words, she had hard-won knowledge plus courageous intention.

For the purposes of this chapter, the main point isn't why Adadevoh had the assumptions that allowed her to lead Nigeria out of the Ebola crisis, but simply that she *had assumptions that differed from those of others*. So do you, and so do I, and both our most mundane and heroic actions result from them; they are equally contained within our own neurological space of possibility. Adadevoh's story illustrates how behaviors emerge out of synaptic pathways to which past experiences have rigged green and red streetlights, so to speak. But what in retrospect was a brilliant response on her part didn't feel "brilliant" or "creative" to her. This is an essential point: it was simply the **most natural** (indeed, possibly the most *personally* rational) idea inside the landscape of possibilities her assumptions had created for her. The space of possibility of Person X (Adadevoh—the person on the left in the figure) **contains** the solution, whereas the space of possibility of Person Y (the Nigerian government—the person on the right in the figure) could not. As such, Person Y is quite literally blind to the solution. This explains a great deal of conflict wherein one person, company, organization, or country simply cannot "perceive" (either literally and/or metaphorically) the action of the other person, company, organization, or country. The issue at play isn't a disagreement as such, but a much

more pervasive kind of pathological blindness that affects many more than those suffering from eye-related pathologies.

Taking the example of the blindness to Ebola's risk of spreading in Nigeria, it is clear, then, that assumptions also have their "other side," or negative possibilities. While they are essential for the brain's functioning, they are not all good (at least in general since everything is necessarily contextual) . . . just like the ideas they produce. They make us vulnerable to useless or even destructive actions, like the officials who nearly let the highly contagious Sawyer out of quarantine . . . or to generalizations across contexts that are not useful. So at the same time that assumptions enable useful perceptions, they also inherently limit them . . . sometimes for good (in that you often don't have bad ideas), and sometimes for bad (in that you often don't have good ideas).

Hence, we are blessed with the fact that inherent in the neural structure of the brain are the processes by which these structures are created. Development never truly ends, as our brains evolved to evolve . . . we are adapted to adapt, to continually "re-define normality," transforming one's space of possibility with new assumptions according to the continual process of trial and error.

For now, to better understand how assumptions limit perception, let's take our theoretical map of potential thoughts for the brain (illustration on page 163) and create a mathematical "God model" of behavior, as it assumes the role of omniscience. We can see everything, not just what happened and happens but everything that could happen or could have happened. This allows us to represent the behavioral/survival value of every possible perception, and we can transform the space of possibility into a *"landscape"* of mountains and valleys. The mountains represent the "best" perceptions and the valleys represent the worst. We've converted our neutral space of *emergent* attractor states, which arise from the interactions of what we in science descriptively call a complex system, into a *fitness landscape*. This idea comes from the study of mathematics, evolution and other fields, including physics, which charts how "fit" a trait (or a "solution" more generally) is, whether in a valley, on a false peak, or on a true peak. When in the valley, the trait reduced survival and the animal/species died. When on a true peak, the animal/species survived better than those on a lower "peak" of the same conceptual landscape. Our thoughts and behaviors have a similar "fitness" in their degree of usefulness to us. In the figure below, valleys are the smallest dots, false peaks are the medium-sized ones, and true peaks are the large ones. Note that they are all generated by the same assumptions.

To know the uber-fitness landscape of life is to know how this landscape changes with time and context . . . and is therefore to know the mind of God. That is the point of any religious text—or indeed any text—that professes to know the truth. The text of science has an attempted rigor to it (say, for instance, Darwin's *Origin of the Species* or Stephen Hawking's *A Brief History of Time*), whereas others do not (e.g., the Bible, the Koran, science fiction). Of course, none of us are God . . . thank God! . . . though some often profess to be, or even worse, profess to know what he/she/it "thinks" . . . in short, to effectively "channel" God. The problem is that in the non-omniscient world of real life, we don't know beforehand which perception is better than another *a priori*. It's the a priori of the previous sentence that's especially significant! Since *that*, of course, is the whole point of trial and error, or empiricism, which is nothing other than a "search strategy" for exploring the landscape of possibility. Like Dr. Adadevoh, we only know what our assumptions make us think, and while we often have confidence about the results we will produce (such as saving lives, in her case), we don't know beforehand what will happen in the future as a consequence . . . despite the irony that the current assumptions are predicted on the success and failure of what *did* happen in the past. To search your space of possibility and find those superior perceptions, like she did, you must draw on the perceptual record of experience that has given you your assumptions. The process of trial and error, whether lived or imagined, is nothing more than exploring this landscape in an attempt to discover the highest mountains while avoiding the valleys. But if there are both good ideas and bad ideas in your space of possibility, what does it look like to land on a good idea?

To answer this, you are represented by the figure in the middle of the space of possibility. This is your current perceptual state. The question is, where are you likely to go to next in thought

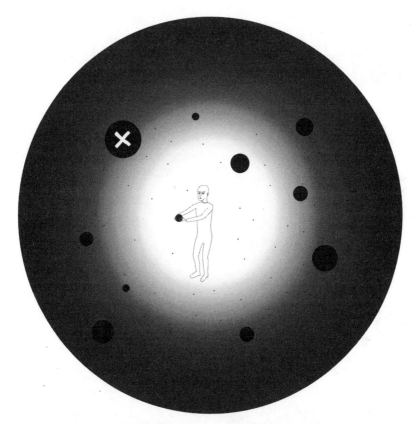

NOT

BRAIN

YOUR

DOES

BIG

MAKE

or behavior? In the case of the development of a flower, Elena Alvarez-Buylla (in collaboration with my lab) showed that the different developmental stages of the flower—carpals to sepals to petals—follow a specific sequence. This suggests that not only were the states themselves selected during evolution, but so was their *actual sequence*. The brain is no different. It too settles into continually changing attractor states of neural patterns, each reflexive response following from another, depending on the nature of the stimulus input given a certain spatial and temporal ecological context.

So, in the above figure, the different possibilities are indicated by the black dots (as noted previously) that have proximities to you. The closer it is, the more likely it is to be the next perception (within "perception" I also include conception, decision, and action). The dots near you in the white zone (the metaphorical "near field") are the most likely "next possible" perceptions . . . that is, the most useful outcomes from the past that have given you your current set of assumptions.

Like evolution itself, the brain only ever makes small future steps to what is the most likely right next step to me given what I did in the past. So, in contrast to the highly motivational concept that "all things are possible," this is not the case *at any moment in time*. Rather, it's a function of an *accumulation* of small steps over time.

Now notice the small X at the top of the dot image in the (invisible) dark zone. While this might be the best idea or decision in principle, because it is at the outer edge of the search space, it is a very unlikely perception (making it effectively invisible). Why? **Because your brain does not make big jumps.**

We can even say that the distant X is even invisible to the person/culture/species at hand. As noted above, this apparently was the case for the officials who at first disagreed with Adadevoh's suggestion to keep Peter Sawyer in quarantine; her good idea was unlikely or impossible for them to arrive at on their own. Their history of experience—past perceptions encoded in their collective brain—put it too far away. Hence, even though there are an infinite number of things you could theoretically think, your past thinking and behavior . . . your assumptions or biases . . . make it more likely that you will go toward certain perceptions and away from others.

So the first *assumption* that I'm trying to encode in your brain through reading this book is this: to admit that you have assumptions (or biases), every second of every day, in every act and perception. At any point in time, we are all only responding, reacting according to our assumptions about inherently uncertain information. Of this you have no control . . . in the moment. This is a good idea much—if not most—of the time. But people who don't know or don't want to admit they have assumptions remain in a state of ignorance of their own brain, and thus of themselves. They live in a state of blindness . . . they can't see the valleys. They only see their own peaks and more peaks until the disaster or failure strikes, at which point circumstances leave them no choice but to discard old bad assumptions and take on new ones . . . sometimes by being selected out of the population itself, hence the infamous Darwin awards. In some cases, however, when disaster strikes whole societies and not only one person, such awards should go not just to individuals, but to institutions.

On September 15th, 2008, the Lehmann Brothers financial services firm filed for bankruptcy after posting billions of dollars of irrecoverable losses, the majority coming from subprime mortgages. These mortgages were housing loans given to people who were unlikely to be able to pay them back; such loans often had scaled interest rates that increased over time and thus became even less viable for the borrowers. Within hours, stock markets across the world experienced vertiginous drops. Soon after, world leaders convened with bankers to discuss measures to stem further damage. In the weeks that followed, the Lehmann brothers bombshell even shifted the focus of the competing presidential campaigns of then-senators Barack Obama and John McCain. Massive layoffs in the financial sector ensued, followed by layoffs in nearly every other sector of the economy. People debated about who was to blame. The beginning of the global financial crisis had arrived.

The simplest way of understanding what happened is that Wall Street made a bad bet. They gambled on the groups of people that lenders preyed on with subprime loans, and this was not a good move. Meanwhile, the governmental institutions that exist to protect our economy (or at least presume to exist for this purpose) did little to stop such reckless betting. What occurred, like most human screwups, was a lived manifestation of bad assumptions. People in power assumed nothing so cataclysmic could happen, as did the banks, who assumed they were "too big to fail." It's as horribly simple as that. They were wrong, yet everyone just kept responding, shaping their environments to fit the biases their brains wanted to confirm. Then the day came when our economy got caught in its own assumptions. The result was a monumental failure to perceive usefully.

As we know, the economic crisis had a disastrous effect on millions of lives worldwide. If policymakers had had different axes of assumptions . . . which would have meant a different space of possibility, with different peaks and valleys . . . perhaps they could have palliated the degree of hardship the crisis and its fallout brought about: the lost homes, broken lives, and dismal poverty. But they didn't, and this is in part due to another facet of our brain's evolution.

As a species, humans evolved in a cooperative context in which survival depended on how well we got along and worked with others, and as we all know, getting along with others is often easier if you don't disagree with them. Because of this, our brain developed a bias toward conformity, and this non-blank-slate quality (assumption) with which we each come into the world affects perception. For example, the rostral cingulate zone of the brain releases chemicals promoting social conformity, or what you could think of as collective kurtosis.[58] This means that other people's spaces of possibilities affect our own, often limiting the perceptions that come naturally to us. This leads us to *memes*.

Here, we can further develop the concept of a meme—originally suggested by Richard Dawkins in 1976—as an assumption of a culture or society that constrains and determines their collective space of possibility (though of course, in Internet-speak memes are the online ephemera that get passed to us daily, often virally). The social response to a "stimulus" in the religious South of the United States, for instance, is going to be different from the social response to the same stimulus in a differently "religious" New York or San Francisco . . . or Japan. To a Japanese person, leaving chopsticks sticking out of rice at an everyday meal is taboo, since it symbolizes death. In Holland it means, well, nothing. Such seemingly insignificant assumptions affect the thoughts we do and don't have, since they form the assumption-axes of our spaces of possibilities. It is quite easy to imagine how more significant cultural/societal assumptions related, say, to gender, ethnicity, or sexual orientation could determine the prejudices certain groups are forced to struggle against, because they are struggling against the collective brain (to which they belong). As such, different memes produce different voting patterns, senses of humor, and values . . . as well as racism. And these memes have stability, as they are physically manifest as attractor states within the brain and are consequently difficult to shift.

Think of the death of Trayvon Martin in 2012, the unarmed African-American 17-year-old wearing a hooded sweatshirt who was killed at night by a gun-wielding Floridian civilian who wrongfully and tragically assumed the boy was dangerous. Or think about all the "legal" murders of innocent black men. Many Western societies have a perceptual bias toward young African-American and black British men, especially those dressed in certain ways. On a cultural level, many have been taught that to fear them is useful by observing the response of others who are afraid . . . even within black communities. The comedian Ian Edwards satirizes this in a bit in which he describes two "black dudes in hoodies" on the same street, each scared of the other. "Hey, man! I don't want no problems, player," says one. The other responds: "I don't want none either, bro . . . *Grandma, is that you?!*" "Fear of other" is a very strong bias **to go away** from one thing and **to go toward** another . . . usually more familiar thing. When we engage in it—as more of us do than we'd like to admit—we are usually acting not only on our own perceptions, but on erroneous inherited ones.

It's important at this point when talking about biases and assumptions and their subjective roots, to mention that here we're **not** talking about post-modern relativism, a type of thinking that gives equal validity to all things by the mere fact of their existence in a fragmented world. Not all perceptions—including those shaped by social history—generated by your brain are equally good. Some are better than others. If this weren't the case, then evolution itself wouldn't have evolved. There would be no such thing as *relative fitness*, since it's the relative nature of fitness that enables change. For example, stoning women and genital mutilation are fundamentally and objectively *bad*, no matter how creatively you justify such behaviors. While one could conceivably argue that these acts are "useful," since the people carrying them out are better able to fit into certain cultures, they are still objectively bad. What is useful is to understand their source, since then there's the possibility of change.

To get a sense of how memes and assumptions more generally can limit your space of possibility when encountering a task or problem that requires creativity, let's do another one of my "reading" exercises. Look at the 15-character string below and combine the letters to form five three-letter words *as fast as possible*. Don't overthink it. Ready, set, go.

A B O D T X L S E M R U N P I

Write down the three-letter words:

1.

2.

3.

4.

5.

Now do the same thing, again as fast as possible but using this string of letters:

L M E B I R T O X D S U A N P

List the words again:

1.

2.

3.

4.

5.

You are likely to have come up with a distinct set of words this time, right? The point here is that what people generate from the first versus the second letter string is typically quite different. In most cases, not one word is the same. Note, however, that the two letter-strings are in fact made up of the same letters, just in different orders. Your assumptions determined the different paths of brain electricity your thoughts took.

So why did you come up with different words? Why did you come up with known words at all? I didn't ask you to. The reason is that your brain assumes that adjacent letters are more likely to belong to each other. This is a natural tendency that has formed in your brain from the assumptions that it developed when you learned to read. What's more, for many their brain has learned to search from right to left. Additionally, I'm guessing you had an assumption of what a "word" was. Did you choose from English? You could have chosen from another language, or even made up words from a nonsense language (we never said that the "words" had to be of a particular kind). But you likely went with the literal and figurative path of least resistance. From the very start, you excluded countless perceptions from your space of possibility. So what was going on inside your brain that made you do this—and, more to the point, *why* did you do this?

A useful metaphor is to think of the brain as a rail system across the UK, or any country with an expansive and well-run railway infrastructure. The train stations are your brain cells, the rails between them the connections between brain cells, and the movement of the trains is the flow of activity across the brain.

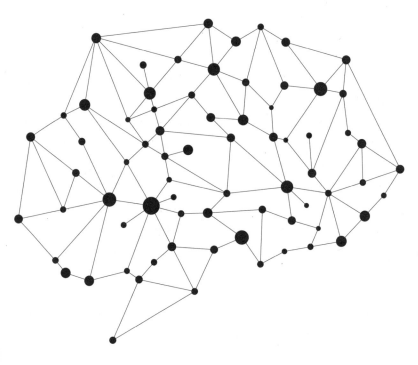

While a network of connectivity is a good and necessary thing, it is by nature limited, since the brain (like a railroad system) is expensive to maintain: though it comprises only 2 percent of your body mass, your brain can consume 20 percent of your energy needs. (For chess grand masters, the process of thinking during a match can "cost" up to 6,000 or 7,000 calories per day.[59]) Your assumptions *are* the connected network of the train system. Your history and past experience of perceptual usefulness constructed them and their necessarily limited routes. After all, no railway infrastructure could sustain the cost of putting track between all possible stops in every locale in the country to offer passengers transport to all the other locales, or even half of them.

In the letter-string exercise, the prearranged, scheduled stops of your neural train kept you from being more "creative" (as viewed in the traditional sense). Your electrical pulses transported your perception on a limited system, thus removing ideas from your space of possibility so you couldn't step to them. Your assumptions about language constrained what was perceivable, though now that you know this, new possibilities suddenly materialize in your brain. If you're up for it, go back and do the exercise again. Your perception will have changed.

It should now be clear, then, that you have the weight of past perceptions holding you where you stand among those dots. Of course, you knew this long before I had you do this "test." We've all lived such experiences: coming up with a perfect comeback hours after being slighted; realizing the apartment you should have gone with after renting a different one; perhaps even being in a relationship/friendship/collaboration with the wrong person. Or simply not having *any* good ideas or solutions for a situation. And we've all heard the expression "hindsight is 20/20," or at least experienced what the idea captures. The Great Recession embodies this, along with all the "Monday morning quarterbacking" that accompanied its fallout. The tragedy of September 11, 2001, embodies it as well. The *9/11 Commission Report* attributed what happened the day terrorists flew two planes into the World Trade Center to "failures of imagination" by the government tasked with protecting American citizens. Before 9/11, however, that failure to imagine was simply the way most of us saw the conceivable possibilities that existed in the world. All of this is the result of the neurobiology of assumptions—assumptions that extend into our bodies—the shape of our hands and ears and eyes, the distribution of the touch receptors across our skin, the biomechanics of movement itself . . . and even into the objects we create. The web of a spider, for instance, is in fact an outgrowth of its assumptions, called an "extended phenotype," where the spider extends itself into the world beyond itself. And take antelopes, which carry the communal brain bias to respond to the visual behavior of the others: if one antelope sees a lion, they all effectively see the lion as if they are *one distributed* perceptual system.

But your past is not your present. The electric patterns in your brain (and the distributed patterns across communicating brains) aren't necessarily the ideal ones just because they might have been the most fit responses "once upon a time." Indeed . . . what was once useful may no longer be useful.

The natural world is in constant flux. Life is correlated noise that matters. If your world is stable, remaining still can be the best strategy. But the world is not stable. It *usually* changes (though not always meaningfully). Evolution is the literal embodiment of this fact, since species that move (evolve) *live* and movement *is* life, which includes relative movement, and thus equally can mean remaining still (or constant) when all about you shifts (as in Kipling's poem "If"). Remember, context is everything. Our traits must stay useful; otherwise we disappear and take our genes with us . . . and the resulting assumptions inherent in the structure of our brains. This, of course, is what happened to our vanished evolutionary cousins the Neanderthals and other hominids. This changefulness of reality is just as evident today. Industries collapse and new ones arise, just like all the jobs within these industries. Likewise, our relationships change . . . with our friends, our families, and our romantic partners. Flux is built into these incredibly important contexts, so we must *flow with the flux*. We must be adaptive . . . the most successful systems are!

In fact, the more the world becomes "connected," the more each event in that world becomes "conditional" (or contextual)

on the events around it in space and time. This is an incredibly important point. We all know the saying "When I was young . . ." Well, the world truly is different now than it was before. Previously, what happened to the Aztecs on any given day, no matter how constructive or destructive, had little immediate influence on the other societies or cultures that coexisted in other parts of the world. That's not the case today. Today, the stock market crashes in Tokyo and the effects are felt in New York even before the traders on the New York Stock Exchange open their eyes to a new "future bear" day. The possibility of new emergent global attractor states is more probable (hence the global financial crisis is a negative example, although of course not all examples are negative: the assumption that we are free to speak our minds, the Internet, and the World Cup are some positive examples). And like all developing, highly connected systems, the resulting attractor states are less predictable every day, a condition that is paralleled by the unpredictability of the weather with climate change.

In short . . . the world's ecology (physical and social, which combine to shape the personal) is becoming more uncertain as the actions of others are now felt much more immediately.

To temper against this, religiosity is increasing, the fear of "otherness" and more generally a fear of losing control (evidenced by the Brexit vote in the UK—or more accurately England). So too, then, are *global* memes. There is an alternative to these latter strategies: Rather than impose an artificial order that doesn't belong, we must change with change, as it is inherent in our ever-transforming ecology. This is a deeply biological solution. It's what we—and other systems—evolved to do. And to explicitly reiterate a previous point (as it's essential to remember), in nature the most successful systems are the most adaptable.

If we don't, our brains will surrender to their previous

momentums and we will only cling to old, *unknown* assumptions, increasing the stubborn stability of the personal and social attractor states since they are simply the ones that are there, and thus deepening the attractor states that inhibit us. Our assumptions make them inevitable.

Or do they?

This brings us to a potentially (and hopefully) disorienting moment in your reading experience—and a very important one. At this point in the book, you may be wondering if you've run face-first into a contradiction that makes my whole claim about the possibility of seeing differently suspect. Earlier we have seen that our brains didn't evolve to see reality, since that would be impossible; and thus they brilliantly "make sense" out of the senseless. Now we have a physiological, brain-based explanation for why only certain perceptions are likely to occur (and why even fewer actually do). Here is where we hit the problem: If everything you do—indeed *who you are*—is grounded in your assumptions; and if your assumptions represent your personal, developmental, evolutionary, and cultural history of *interacting* with your external *and* internal environments (i.e., what I call your ecology); and if these assumptions produce reflexive responses over which you have little—if any—control *in the moment*, then how can you ever break out of this cycle and see differently? Aren't we trapped forever in fixed sequences of reflex arcs of seeing what we saw before, turning us into legs that kick automatically (perceptions) every time the doctor (stimuli) taps us in the same spot? Shouldn't it be impossible for your brain to do anything but use the same old neuroelectrical train routes?

Up to this point, *Deviate* has created the space for you to see yourself see . . . to become an observer of your own observations, a perceiver of your perceptions. We have learned that our

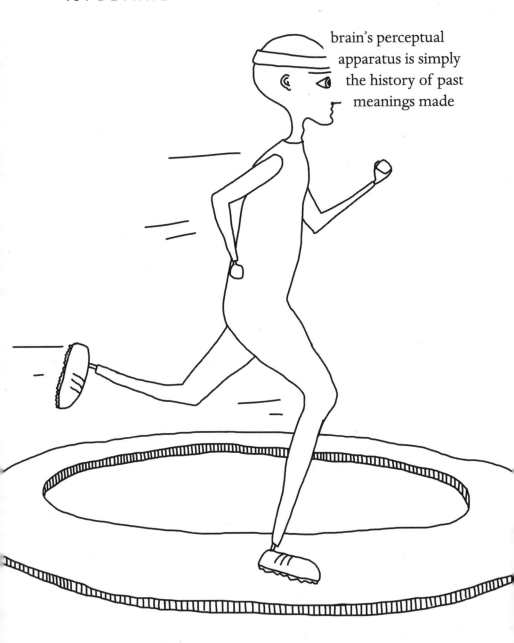

brain's perceptual
apparatus is simply
the history of past
meanings made

physically manifest. But more than this, we also learned that the process of creating assumptions is just that . . . an inherent process of the brain's constructive process itself. Therein lies our "true salvation." The process of constructing perceptions is the route to both constraining what we perceive and also changing it, if not expanding it.

You have already accomplished the first step in deviating thoughtfully. You are aware that you are usually unaware.

Which means . . . a bit confrontationally, though I mean it more challengingly . . . now that you know this, each of us no longer has the excuse of ignorance, which is too often the initial barrier to change. If I've replaced your brain's default assumption of "knowing reality," then my aim so far has been achieved: you now know less than you did before. And in knowing less, you (and we) now have the opportunity to understand more. Without this step of understanding, every decision you make in the future will remain a response grounded in history . . . for better or worse. There will not be any choice in the matter, despite the narrative your brain tells itself. Choices exist where there is an option. Understanding why you see what you do provides that option. It gives you the potential of choice, and therein the *potential* for intention.

Seeing differently—to deviate—begins with awareness . . . with seeing yourself see (but by no means ends there). It *begins* with knowing that some of those often invisible assumptions that maintained your survival in the past may *You are aware that you are usually unaware.* no longer be useful. It begins with understanding that they may in fact be (or become) bad for you (and others), and if not changed would curtail living. To truly embody this is to empathize with what it is to be human . . . indeed, with what it is to be any living, perceptual system.

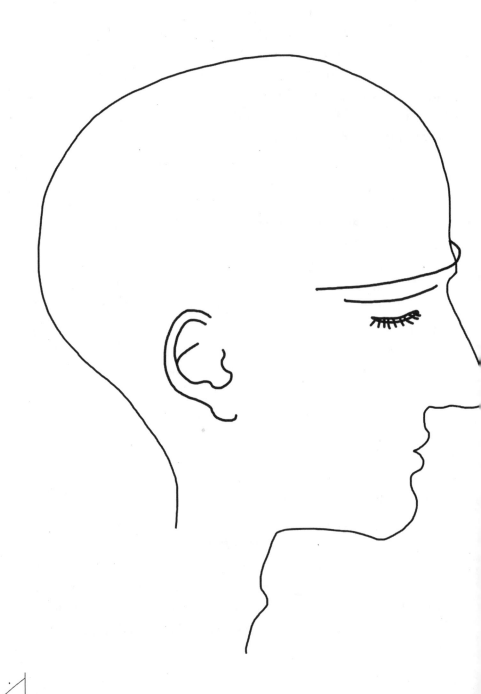

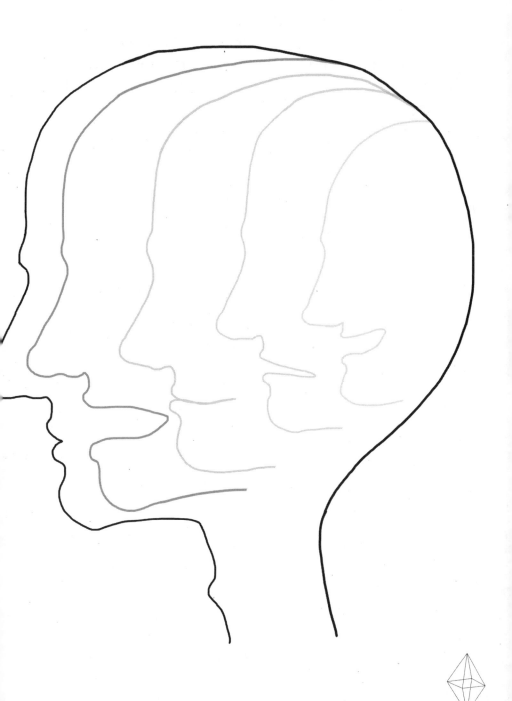

So how do we see differently?

We change our future by **changing our past**.

As strange as it may sound, this is entirely possible. As a matter of fact, it's what we do all the time. Every story, every book, all narratives spoken, read, or enacted are about changing the past, about "re-meaning" past experiences, or more specifically, about changing *the future past*.

CHAPTER 7

Changing the Future Past

One of the most famous and controversial experiments in the history of neuroscience, done by a man named Benjamin Libet in the early 1980s, had a very simple premise: participants had to move their left wrist or their right wrist.

A researcher in the Department of Physiology at the University of California, San Francisco, Libet died in 2007 at the age of 91, but his 1983 paper that came out of the experiment is still legendary. He discovered that there is a time lapse between the behavioral decisions our neural circuitry makes for us and our awareness of these decisions. His results set off a debate about the brain, human consciousness, and free will that is still raging today. Why? Because his findings challenge the very idea that we have agency in engendering new and creative thoughts. In other words, his experiments suggested that we aren't the conscious owners of our destinies . . . we just watch them happen, falsely thinking we're in control. But we do have agency. However, in order to know how to exercise it, we must first understand why it is possible.

Libet's experiment went like this: He and his team began by

fixing electrodes to the scalps of participants to measure the electrical activity in their brains. Then the participants were asked to move either their right or left wrist, but before doing so they indicated the exact instant in which they made their decision about which to move. This was made possible by an ingenious stopwatch-like device, which measured three different things down to the millisecond: the instant the participants' neuroelectric signals indicated that their decision was made inside the brain (the *Bereitschaftspotential* in German, or "readiness potential"), the instant they consciously made the decision, and the instant of the actual movement of the participants' wrists. The results? On average, the *Bereitschaftspotential* in the participants' cortexes came 400 milliseconds *before* their self-aware decision to move, which in turn came 200 milliseconds before the actual movement. While this finding may seem straightforward enough, suggesting a natural "sequence," the philosophical implications of the experiment were—and *are*—profound.[60]

In Libet's (and many others') interpretation, his discovery revealed that the participants' conscious decisions were a fiction. They weren't decisions at all . . . at least not as we commonly think of them, since they happened in the brain before the participants were conscious of them. The relevant attractor state of activity within their specific brain networks was present before their conscious, nominally decision-making minds. Only after the fact did the decision appear in their consciousness, masquerading as the cause of the movement. By implication, this means that decisions in the present don't necessarily belong to conscious proactive intentions as such, but to the neural mechanics that determine automatic perceptual behavior. By extension, it suggests that *free will doesn't exist*. If correct, Libet's experiment would mean that humans are passive spectators to the ultimate virtual reality experience: their own lives.

Over the years, Libet's findings proved to be so provocative that they spawned a whole new discipline of research ... the neuroscience of free will. His experiment also upset and pleased philosophers in equal measure, depending on where they fell in the ancient argument of determinism versus free will, since his experiment confirms that we do not have control over what we do now since everything we do *in the moment* is a reflexive response, even though it doesn't feel that way. We only ever react in the here and now ... at least when we are *unaware*.

But a lack of proaction does not mean we can't act with intention. The key to unlocking this process of acting with intention is awareness. Once we are aware of the fundamental principles of perception, we can use the fact that we don't see reality to our advantage. To do so, we must remember that all of our perceptions represent nothing other than our and our society's past perceptions about what was useful (or not). So while we're not able to consciously control the "present *now*," we can influence our "future *now*." How? By **changing our future past**, which raises a deep question about where free will—if we have it— might actually live.

What do I mean?

Libet's experiments demonstrate that we have little ... if any ... free will over our responses to events *in the present*. But through the process of imagination (delusion), we do have the ability to change the meanings of *past* events, whether they occurred a second ago or, as in the case of some cultural memes, centuries ago. "Re-meaning," or changing the meaning of past events, necessarily changes our "past" history of our experience of the world—not of course the events themselves, nor the sensory data arising from those events, but the statistical history upon which perception is predicated. From the perspective of perception, exercising the free will to re-mean the past history of meanings (i.e., our narrative) changes our *future* history from that moment on . . . hence, our "future past." And because future perceptions—just like the ones you're experiencing now—will also be reflexive responses to *their empirical* history, changing our "future past" has the potential to change *future* perceptions (each, ironically, is generated without free will). Hence, almost every story we construct about ourselves in relation to the world, whether they are born out of consultation with a psychotherapist or from behavioral cognitive therapy or from reading a "pop-sci" book like this one, is an attempt to re-mean past experiences in order to change future reflexive behaviors individually and/or collectively.

But HOW do we begin to change our future past **in practice?** Answer: By starting with a question . . . or with a joke.

The great Czech writer Milan Kundera's first novel, *The Joke*, is a perfect—and perfectly layered—example of just this. The central character is a young man named Ludvik who makes a joke that turns out to be the wrong one for communist Czechoslovakia in the 1950s, a time in which "fun went over badly."[61] He writes a postcard to a girl he has a crush on, who he doesn't feel appreciates him. It reads: *"Optimism is the opium of the people! A healthy atmosphere stinks of stupidity! Long live Trotsky!"* She shares his subversive missive with the authorities and this horribly recasts his future, causing him to commit a cruel deed many years later. But at the book's end, a matured Ludvik reflects on his past, arriving at a deterministic—and perhaps convenient—conclusion. He decides that the effect of the joke he made, as well as other seemingly harmless acts, are the result of historical forces beyond human control (a clear argument against free will): "I suddenly had the feeling that one's destiny is often complete long before death."

The irony is that not only does *The Joke* tell the story of upheaval in Ludvik's life, but the real-life story of the book's publication led to upheaval in Kundera's own life and in the life of his country. Soon after *The Joke* was published, the radical social uprising of the Prague Spring of 1968 embraced it, absorbing its irreverent attitude into a heady rebellion against the repressive government, which promptly banned the novel. Kundera's book, like Ludvik's joke, went on "monstrously multiplying itself into more and more silly jokes." Soon after, Kundera lost his teaching job and entered exile in France, changing the course of his life. The authoritarian regime saw the novel and its titular joke as a threat, and its author as a threatening *deviant*. This is because governments—especially totalitarian ones—and their spin doctors understand the power of re-meaning history. Those who influence the meaning of the past will shape the foundations by which those who identify with that past behave in the future. Hence asking *Why?* about the past on something as objectively harmless as ink and paper, as did Kundera's novel and like the work of so many before and after him, became an act of rebellion that rippled in significance . . . which ultimately shaped his own future. Years later, Kundera quipped in an interview that every one of his novels could be titled *The Joke*.[62]

All of this happened because Kundera didn't simply publish a work of fiction but also engaged in the most dangerous thing a person can do and has done throughout history. He asked a question: why?

To ask why is evidence of awareness . . . of proactive doubt. And *The Joke* is evidence of the power of why. In particular, the subversive quality of **Why?** is to be found in the change that it has created throughout history, and in its active suppression by governments and institutions, religions, and—most ironically of all—educational systems. Hence, innovators begin the process of creating new perceptions—of changing their future past—by asking why not just of anything, but of **what we assume to be true already** . . . *our assumptions*. Arguably, to question your deep assumptions, especially those that define you (or your relationships or society), is the most "dangerous" thing you can do, as it has the greatest potential to lead to transformation and destruction in "equal" measure. It can have such a seismic impact precisely because it recasts the past, giving you new ways of thinking about concepts and circumstances that had previously appeared to be simply a set reality. If you don't ask why you have one response, there's no chance of creating a different one. Yet learning to constantly ask why is not easy, especially in a time when information is seen to be so essential.

"Big Data" is the buzzword phrase of the early twenty-first century that is as powerful as currency. An obsession with it has taken root in many areas of our societies, from medicine to commerce, all the way to individuals in the measuring of their "steps" throughout the day. There is even a popular band called Big Data. The term simply refers to sets of data so immense that they require new methods of mathematical analysis, and numerous servers. Big Data—and, more accurately, the capacity to collect it—has changed the way companies conduct business and governments look at problems, since the belief wildly trumpeted in the media is that this vast repository of information will yield deep insights that were previously out of reach. By gathering metadata on our behavior . . . which presently is largely restricted to our Internet-based habits of viewing/buying/traveling . . . companies will be able to market to me directly: to my "preferences" (read: assumptions). Netflix *might* be better at recommending movies and shows that are specific to what I like; Amazon *might* better market to me according to what I like to buy in spring, and thus be more likely to make a sale; traffic apps *might* better route me through traffic according to my preferred trade-off between time and aesthetic; and health researchers *might* better locate dangers in my body.

The irony is that Big Data by *itself* doesn't yield insights, since what is being collected is information on who / what / where / when: on how many people clicked or searched something, and when and from where they did so, as well as sundry other quantifiable particulars. All that this bank of who-what-where-when gives you is precisely what the term "Big Data" so nakedly admits: Lots of fucking data . . .

Without the brain (and in the future, brains augmented with AI) to generalize usefulness across contexts—to effectively find wise metaphors that transcend a situation—information doesn't serve us. Without knowing why, we can't find laws (basic principles) that can be generalized . . . like the law of gravity, which applies not to any particular object but to all objects that have mass. Effects without an understanding of the causes behind them, on the other hand, are just bunches of data points floating in the ether, offering nothing useful by themselves. Big Data is information, equivalent to the patterns of light that fall onto the eye. Big Data is like the history of stimuli that our eyes have responded to. And as we discussed earlier, stimuli are themselves meaningless because they could mean anything. The same is true for Big Data, *unless* something transformative is brought to all those data sets . . . understanding.

Understanding reduces the complexity of data by collapsing the dimensionality of information to a lower set of known variables. Imagine you're part of a startup that has developed a new heating device of some sort, which you want to market in a targeted manner. For your research, you take a series of measurements of the body temperature of living animals, and in particular the rate at which they lose heat. You find that they all lose heat at different rates. The more animals you measure—including humans—the more data you have. Given your dedication and diligence in measuring, you have accrued a huge dataset, one that is of an increasing number of high dimensions, in which each animal is its own dimension, despite the seeming straightforwardness of the simple measurement. But the measurements themselves don't tell you anything about how or why there is variability in heat loss among different animals.

What you want is to organize the data set. But in principle there are a huge number of ways to do so. Should you organize it by type, color, surface quality, or a combination of one, two, or n variables? Which is the best (or "right") way to organize it? The "right" answer is the one that offers the deepest understanding, and it turns out that in this example the answer is by size. We know—because someone did exactly this experiment—that there is an inverse relationship between size and surface area: the smaller the critter, the more surface area they have proportionally, and the more heat they lose and the more they need to compensate for that heat loss in other ways, thus creating the conditions for the trial-and-error process of evolution to find the solution.

There you have it: a generalizable principle. What was once a massive, high-dimensional dataset has now collapsed to a single dimension, a simple principle that comes from using the data but is not the data itself. Understanding transcends context, since the different contexts collapse according to their previously unknown similarity, which the principle contains. That is what understanding does. And you actually feel it in your brain when it happens. Your "cognitive load" decreases, your level of stress and anxiety decrease, and your emotional state improves.

To return to dour, put-upon Ludvik, does his philosophy on life apply to human perception? Is your perceptual "destiny"

already "complete," out of your control because it was shaped by historical, evolutionary forces that preclude free will? Absolutely not. "Why?" provoked not only the Prague Spring but also the French Revolution, the American Revolution, and the fall of the Berlin Wall. The revolutionaries and everyday citizens who brought on these societal waves of change all shared the same question: *Why are things this way and not another way?* If you get enough people to ask themselves this, tremendous—and tremendously unpredictable—things suddenly become possible (without being able to define what those things may be *a priori*). And the reason is simple: who, what, where, and when lead to answers that are lit by the metaphorical streetlamp illuminating the space that we can see (i.e., *measure*). Of course, measurement is essential, as are descriptions more generally. But data is not understanding. For instance, while traditional schools continue to teach what is measurable (e.g., the kinds of answers that come through rote learning), such methods do not teach understanding in the children being measured. It's teaching under the streetlamp, when we know that we dropped our keys somewhere else, in a place of darkness. Rather than search in the dark, we stay in the light and harvest more and more measurable data. Notwithstanding incredible feats of engineering technology required for some kinds of measurement, gathering data is easy. *Understanding* why is hard. And, to reemphasize the point . . . the cash isn't in knowing, it's *in understanding*. Thus, when thinking of the rise of TED Talks (the popular online videos of interesting thinkers explaining their ideas onstage to an audience), rather than "ideas worth spreading" we need to consider **"questions worth asking."** Good questions (most aren't) reveal and build links in the same way the brain does in order to construct a reality . . . a past we use to perceive in the future . . . out of the objective one that we don't have access to.

This is why George Orwell was so wise when he wrote, "Every joke is a tiny revolution."[63] It's worth noting that asking *why* follows a long tradition . . . the tradition of philosophical thinkers since time immemorial, from Socrates to Wittgenstein. What philosophers do is take previous assumptions (or biases or frames of reference) and question them. This leads them to elaborate on them or tweak them or try to destroy them, to then replace them with a new set of assumptions that another philosopher will engage with in the same way. This seemingly esoteric methodology of questioning is anything but esoteric. It is not only a learnable skill, it is of incredible practical importance in a world obsessed with finding definitive, cure-all answers. This is why we must bring back the craft and way of being of philosophy in our everyday lives. Anything creative is initiated by this kind of philosophical questioning, which is why it might be one of the most practical of our dying intellectual disciplines. Schools rarely even teach children how to ask questions, much less what a good question *is*, or the craft of finding it. As a result, we're—metaphorically and in a way quite literally—"great engineers but crap philosophers."

Questioning our assumptions is what provokes revolutions, be they tiny or vast, technological or social. Studying the brain has shown me that creativity is in fact not "creative" at all, and that, at heart, "genius" emerges from simply questioning the right assumption in a powerful, novel way. The story of the Rosetta Stone demonstrates this.

Discovered by a French soldier in 1799 near the small port city of Rashid on the Nile, the Rosetta Stone was a charcoal-dark slab of granite measuring nearly four feet high and more than two feet across. Repurposed from a disappeared building (possibly a temple) to form part of a fort, it was etched from top to bottom with intricate inscriptions in three different "scripts": ancient

Greek, hieroglyphs (a type of "high" formal writing used by priests) and demotic (a "low" common writing). Hieroglyphics and demotic were two different ways to represent the ancient Egyptian language, though no one knew this yet; languages and orthographies had mixed and mutated across centuries and empires, leaving a confused and confusing trail of clues about their evolutions. The potentially revolutionary importance of the stone was immediately apparent to scholars, who knew relatively little about the pharaonic civilization that had left behind the pyramids at Giza with their tombs, mummies, and other enigmatic artifacts . . . and knew *nothing* of their language. As a result, the stone seemed a godsend, offering the most direct access to these writing systems and thus to their long-gone underlying civilization. To use a cool, geeky term of today, the Rosetta Stone might be a "cryptanalytic crib."

The discovery of the stone took place at a time of international military intrigue. Napoleon had invaded Egypt in 1798 with the intention of attacking British hegemony by taking control of this far corner of their empire. He failed, but this incursion prefigured and even determined what was to come for the Rosetta. It caused the British and French to both be present at the discovery of the stone—it might not have ever been discovered if not for the French, but the victorious Brits took it home. This set the stage for a battle of scholarship between the two countries, like a mirror of their warring imperial ambitions. Since the scholars at the time knew Greek, they thought it would simply be a matter of translating the stone into the other two "languages," thus deciphering the code and subsequently all other writings left behind by ancient Egypt. The only problem was that no one could figure out *how* to use the Greek to decode the hieroglyphics and demotic—at least initially, until Jean-François Champollion.

Born in 1790, the French Champollion was still a boy when the Rosetta Stone was found, though he would eventually come to be revered as the father of Egyptology. The son of a bookseller father and an illiterate mother, by the age of sixteen he had learned over a dozen languages. He was gifted, but impatient with the banal drudgery of school. The one thing that excited him over everything else, however, was ancient Egypt. In his book *Cracking the Egyptian Code*, British writer Andrew Robinson included a 1808 letter from Champollion to his parents that says it all: "I wish to make of this ancient nation a thorough and continual study. The enthusiasm with which the description of its vast monuments transports me, the admiration with which its power and its knowledge fill me, are sure to increase with the new ideas that I shall acquire. Of all the peoples whom I admire the most, I shall confess to you that not one of them outweighs the Egyptians in my heart!"[64]

By his early twenties Champollion had established himself as a respected philologist and scholar (by age nineteen he was a professor). He pushed himself incredibly hard, driven by a feverish desire to learn every language that he could and study every artifact the conquests of Napoleon had appropriated and brought back to Paris. By then small "dents" had been made in the tantalizing hunk of rock found in Egypt, and foremost among the people who had worked on decoding it was Thomas Young, a brilliant British doctor and physicist.

The Rosetta Stone with numbers is a hidden game: if you take an old mobile phone and text 3384283 with the T9 technology, you will write the word: "DEVIATE."

Although Young has largely been overshadowed in history by Champollion, he had a polymathic, Goethe-like wingspan of interests of his own, but unlike Goethe, he was a truly formidable man of science. He too studied the perception of color and in fact made one of its most fundamental predictions—that the human eye uses three types of receptors in daylight (called trichromacy), despite himself being "color blind." Young had expended immense time and effort in studying the Stone's puzzling inscriptions but hadn't landed on an insight that would unlock them. When Champollion returned to his attempt at translating the Rosetta in earnest in 1815, an arms race of scholarship began.

Young and others had been working from the assumption that the Egyptian hieroglyphics were only symbolic—characters that represented concepts, with no correspondence to a spoken language. But in January 1822, Champollion obtained a copy of a newly discovered inscription from an obelisk related to Cleopatra, which allowed him to better understand the complexities of hieroglyphics. One of these complexities was that phonetic signs were used in ancient Egypt.

From this, for Champollion to unlock the Rosetta Stone he needed to question Young's baseline assumption in order to institute a new, correct one: that the hieroglyphs *did* represent phonetic sounds . . . words from the Coptic language, an outgrowth of ancient Egyptian, which Champollion just so happened to speak from all his tireless work slurping up languages. He was so overwhelmed when he had this breakthrough that he supposedly passed out after shouting, "I've got it!"

We typically consider creativity the "Aha!" moment of bringing two things together that were once *far apart*. We think, "Wow! How did you bring those two things together like that? I never would have thought of it!" The seemingly further apart the ideas, the greater is the genius. We assume big causes lead to big effects. As the Rosetta Stone shows us, to understand what is really going on we must sweep away the myths to examine the perceptual mechanics.

Let's experience firsthand the power of changing assumptions—through questioning them—by returning to the diamond whose directionality we changed in Chapter 5. Before you flip the book and "make" the diamond spin, your brain is in a particular physiological (and thus perceptual) state . . . namely, of seeing a set of stationary lines. We can represent that "state" by referring back to our egocentric circle diagram in Chapter 6. You are represented as the "person" in the middle of the diagram.

Now . . . start flipping the book.

Because your brain has an inherited bias (assumption) that we typically look down onto planes, the most likely direction of rotation, given your history of experience, is to the right. So when you flip the book, even though the lines are not actually moving, your brain takes the small physical differences between each image and perceives the probable meaning of the information—not the information itself—and sees the object to be in motion, specifically spinning to the right; it depends only on your brain's subjective interpretation, which draws on your past history of usefulness. This reminds us that not all perceptions are equally possible at a given moment. Some are always more likely than others.

I've represented the most likely steps of what you perceive according to distance to your current "location," and rightward rotation is close to the center. Your brain made a small step based on its history of interaction with the world.

So, while there might be other perceptions that are theoretically possible (the diamond spinning from right to left), it's in the "dark

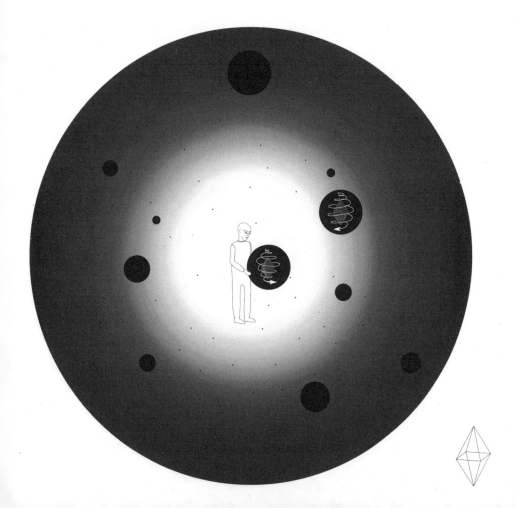

zone" of what is possible. But change your assumptions (that you're looking up—not down—on the surface of the diamond), and your space of possibility changes too, enabling previously impossible perceptions to become possible (and vice versa).

Remember, perception doesn't exist in the *Star Trek* universe: behaviors don't teleport. Your heart rate will increase gradually, not jump suddenly from 50 bpm to 150 bpm without passing a rate in between. Your muscles can't cause you to be sitting one instant, then standing the next. Champollion didn't happen upon the beginning of Egyptology. A host of tiny linear movements occur. As with the body, so with the mind: you don't get beamed

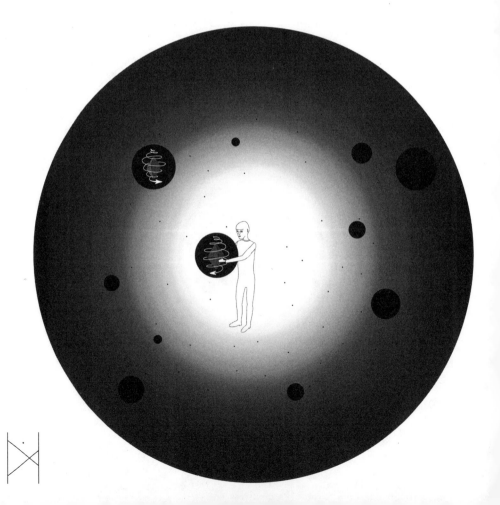

to big ideas. The process of "coming up with ideas" may feel like leaps, but thousands of minute, sequential processes actually take place. Just as the development of a flower goes from carpals to sepals to petals, you don't just suddenly soar across your space of possibility; you are constrained by your circuitry that represents what was "best" *in the past*. Let's look at a very contemporary example.

In March 2008, Apple released the iPhone. Across the world people waited for hours and hours in long lines to buy the new device. It is reportedly the best-selling product in the history of humanity (rivaling the Bible). But a tally of units sold and money made isn't interesting or illuminating; this is information that describes who, what, when, where, or how (or how much). What is important is to look at how the iPhone and subsequent smartphones changed us, and **why** they did so.

Smartphones gave us a new way to interact with the world and with each other (though one must not simply glorify them for the sake of it, acknowledging that this "new" way of interacting is in *some* instances a worse way). They closed the gap between our longstanding physical lives and our now-parallel digital lives, giving us the best bridge yet for bringing these two realities into harmony to make them just one life, which for better or worse is the future. This is why smartphones have changed the way we live, and why Steve Jobs and Jony Ive, Apple's chief designer (as well as many others at Apple), have affected so many lives, including yours, whether or not you own one of their products. They offered an answer to a question that transcended geography, culture, language, and individual personalities. That is a truly stunning achievement, to tap so deeply into universal human experience . . . an experience that is above all else defined by the brain and the perceptions it creates.

But what is critical to understand is that the steps taken to create such a massive shift were small, each initiated by a question ... in particular the question *why*. Apple didn't magically teleport to the iPhone. Jobs, Ive, and the whole company had "why-asking" past patterns of perception, which made ideas that would seem out-of-the-box or literally unthinkable to another person feel intuitively logical to them. They were naturally occurring thoughts inside the Apple-specific box, though this doesn't mean they didn't put intense effort and work into finding them. To bring back our neuro-attractor model for perception, the assumptions the Apple team had acquired let their collective brain-train make stops at ideas others didn't even know could be stops. Each question changed the structure of the space of possibility, sometimes massively so, and not only for one person but for many, many people. Because of it, the space of possibility for others is now forever changed.

A very tricky nuance when it comes to the concept of genius and revolution is that once we share our ideas with the world at large it's impossible to know what will happen. Some insights, like hieroglyphs, represent spoken language; or concepts such as Kepler's discovery that we aren't in fact the center; or even products like the triangular sail that enabled sailors to move upwind for the first time, or Steve Jobs's vision and Jony Ive's design philosophy at Apple, have a meteoric impact on lives and people. Others, like the 1980s LaserDisc "innovation" in home cinema, go plop and disappear. So why is it that some ideas or questions lead to big changes, while others do not? Why did Milan Kundera have to go into exile after the Prague Spring instead of some other Czechoslovakian novelist who also published a novel at the same time? Why did Steve Jobs change modern life forever and I didn't, or you didn't? Is there a way to explain this? Yes, there is.

A large person jumping into a pool makes a large splash. A

small person doing the same cannonball makes a smaller splash. This is obvious. Ideas, however, don't obey the same rules, which is why innovative perception is so powerful . . . and so desired.

To understand "genius," we must challenge the assumption that big effects must have arisen from big causes. They don't have to, and arguably the best ones don't. Everyone seeks the small investment that leads to "unicorns." The answer is a question, which while seemingly "small" has massive effects; this, I'm suggesting, is what defines a "good" question. Whereas big-seeming questions (ideas) can have little or no effect, mathematics has been studying this observable truth, and the explanation takes us into one of the most ground-moving realms of science of recent years . . . complexity theory. By understanding why some ideas cause massive change, while others don't, we can learn to more effectively construct and target questions that produce so-called genius breakthroughs. Why questions are those that will lead to breakthroughs, especially when why is challenging a known truth. But because we are so entrained in assuming linear causality, and thus assume that creativity arises from linking disparate elements through spontaneous insight, it is essential to understand why this is the case. And in doing so, this awareness can lead to new ways of seeing, experimenting, and collaborating.

Complexity theory is a general area of scientific research that has emerged in the last thirty years, and an area of research that my lab has been actively engaged in . . . in the context of adaptive networks . . . for fifteen years. It helps us to study phenomena both natural and man-made, from earthquakes and avalanches to economic crises and revolutions. Complexity theory has helped scientists prove that such influential events are actually systems of interacting parts that follow mathematical patterns and principles. Upheavals are impossible to foresee, but after the

fact, their "upheavability" falls into a neat breakdown, or "power law." This is because complex systems can eventually progress to a "self-criticality" where they occupy a fantastically thin threshold between stability and convulsion in which the falling of a grain of sand (or the publication of a novel in Czechoslovakia in 1967) may cause absolutely nothing, or unleash a landslide. And the brain is the most complex living system . . . if not the most complex system altogether . . . in the known universe.

The basis of complex systems is actually quite simple (and this is not an attempt to be paradoxical, like an art critic who describes a sculpture as "big yet small"). What makes a system unpredictable and thus nonlinear (which includes you and your perceptual process, or the process of making collective decisions) is that the components making up the system are interconnected. Because the components interact, the behavior of any one of the system's elements is affected by the behaviors of all the others to which it is connected. We all embody complex systems because we as humans *are* one, as is the society and larger environment in which we live, including its history of evolution. Yet for centuries, science has pursued a different kind of assumption, of a world that doesn't actually exist . . . a world of "linear causality," largely because it seems the brain evolved with this assumption as a useful approximation. It is a place where A causes B causes C, epitomized in the formulas of Newtonian physics where all the world is not a stage but a billiard table. The Newtonian view works *approximately*. The energy from this causes the movement of that; happy cause results in happy effect. This model is useful, but not accurate. The problem is that breaking a system down into its component elements strips the world of what makes it not only natural, but beautiful . . . and fun—*the interaction between the stuff!* Life and the perception of life. Live . . . in the space . . . between.

(This suggests the biologically inspired mantra "Be in between.")

We can think of these interconnected things, whatever they happen to be, as *nodes* (the elements) in a network that usually have non-uniform connections (called *edges*). In any network, some nodes have more edges (connections) than other nodes. This takes us back to questions, but in particular *good* questions . . . questions targeting assumptions.

Life and the perception of life. Live . . . in the space . . . between.

Our assumptions are un*question*ably interconnected. They are nodes with connections (edges) to other nodes. The more foundational the assumption, the more strongly connected it is. What I'm suggesting is that our assumptions and the highly sensitive network of responses, perceptions, behaviors, thoughts, and ideas they create and interact with are a complex system. One of the most basic features of such a network is that when you move or disrupt one thing that is strongly connected, you don't just affect that one thing, you affect *all* the other things that are connected to it. Hence small causes can have massive effects (but they don't have to, and usually don't actually). In a system of high tension, simple questions targeting basic assumptions have the potential to transform perception in radical and unpredictable ways.

Take, for instance, the three simple dot diagrams shown overleaf. They all have 16 assumptions (nodes). In the first diagram, none of the assumptions are connected (there are no edges). In the second diagram, there are some edges; and in the third diagram they are all connected. Now imagine "moving" a single node in the upper right corner by questioning it. Only it moves. Now imagine doing the same thing to the same node in the second image: three nodes move. Finally, imagine doing the same thing to the same node in the third image: they all move.

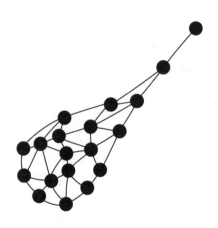

You could think of this as the visual story of a "genius idea." One key insight or realization sets off a chain reaction of other insights and realizations, creating a "house of cards" effect of no-longer-useful assumptions collapsing.

The interplay of assumptions in your brain and in your creative process is what we call an *embedded hierarchy*. They are like your body. Your body is made up of organs, your organs are made of cells, your cells are made up of *organelles* (the organs of your cells), your organelles are made up of molecules, and so on. In essence, a society is nothing other than the changing density of the molecular elements that make us all up. Each is just an attractor state within a larger system, like waves rising on the sea before crashing on the shore, or transient whirlpools on a fast-flowing river. The lower and more foundational the assumption, the more its change can affect the rest of the system, since the hierarchy builds off of it. So, the right question—though small—can cause a person, an invention, an idea, an institution, or even a whole culture to change (hopefully for the better, but not always).

The principle for deviation here is this: *To question begins a "quest,"* and a journey into the unknown. The most insightful quest begins with "Why?" . . . especially when targeted at what you assume to be true already. Because your truths are highly connected assumptions, change them and you may change the whole system. This will mean that the next step you take into that new space of possibility will be "creative." Here choice begins, and in making this choice, you begin the process of changing your future.

The importance of interconnectedness is reflected in the brain itself. In a creatively engaged mind there are higher levels of cortical activity . . . that is, more ideas generating heat both during moments of "inspiration" and in general. This is because the patterns become more widely distributed and thus more connected and interactive. Inhibition also goes down, which means the voice in your mind that loves to shout the Physics of No speaks less often and less loudly. In case you were wondering, some psychedelic drugs like those found in "magic mushrooms" have a similar effect on the brain. Recent functional imaging research by Enzo Tagliazucchi, Leor Roseman, and their colleagues at Imperial College in London has confirmed this to be true. Normally, when we look at a scene, most of the neural activity is restricted to the primary visual cortex at the back of your head. When, however, subjects looked at the same visual scene after taking LSD, not only was the visual cortex more active, but many more areas also became active. In other words, LSD enabled activity to pass between more areas of the neural network, thereby linking more diverse parts of the brain. In line with this physiological change, many people also reported a sense of "ego dissolution" and "oneness." Maybe not surprisingly, then, given this more "open" perspective, low doses of psychedelics (such as psilocybin in "magic" mushrooms and LSD) have been shown to facilitate human relationships, support couples therapy, and lower levels of depression for an enduring amount of time. By connecting more diverse areas of the brain, it's as if these chemicals enable people to see beyond their current entrenched assumptions and create new assumptions . . . expanding their space of possibility, and in doing so altering their future past, and thus their future perceptions (frequently for the good; the response varies depending on the dose taken).

While this might sound like an endorsement of drug use, it's just "information." And if you find it emotionally challenging—as people often do—it's not that the information is itself challenging, since it is meaningless in and of itself. If you find this happening to you, then it's likely because the information is inconsistent with your own set of assumptions, against *your* meaning of the information. But information is agnostic. Meaning isn't. The meaning you make with this knowledge is up to you.

All this is to say that creativity is in fact a very basic, accessible process. It's all about changing your space of possibility by questioning the assumptions that delineate its dimensions. It is, therefore, misleading to divide (or self-identify) the world into "creative" and "non-creative" people. In a dynamic world, we're all hungry fish needing to thrive . . . which means we all need to deviate and therefore start with questions rather than answers, in particular the question why. And this process of questioning assumptions is entirely within reach. We all have spaces of possibility, and live in a dynamic, contextual world, so we can—and indeed must—all change those spaces. In our contemporary society we say that a creative person is creative because they are able to see a connection between disparate things that we were not able to see. But for that "creative" person, the two ideas/perceptions were not a big step belonging to a perceptual process that we physiologically just don't have. Instead, the step was a small one that resulted in a massive change. Considering this, suddenly creativity as we traditionally think of it is not creative at all. Creativity is only creative from the outside.

In other words, if we just learn how to use the tools we already have inside our brains, to change our future past with our own well-aimed whys, this is how we change the *Creativity is only creative from the outside.* "destiny" that Kundera's Ludvik couldn't alter. But if there's nothing creative about creativity (as typically defined), what makes creativity so difficult?

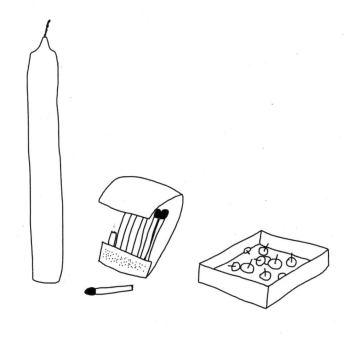

CHAPTER 8

Making the Invisible Visible

Revolutionary questions and the revolutions they start come from demolishing old assumptions to institute new, wiser ones. The problem is we are often blind to why we do what we do. As we have seen, the assumptions that shape our perceptions are often like the air that sustains us . . . invisible, which makes it hard to know where to ask and target our why questions. In this chapter we will learn to see what is unseen.

Have a look at the objects in the image on the previous page: a candle, a matchbook, and a box of thumbtacks. Your task is simple: place the candle on the wall and light it using the items in the image, which is called "Dunker's candle problem" after its creator Karl Dunker (please don't look ahead to find the answer). Note that the thumbtacks in the box are not long enough to penetrate through the candle itself.

Are you struggling? Most people do, as did I when I first saw this. Despite the "right" answer (the X in the space of possibility) being objectively easy, most people take a long time to find the solution, if they find it at all. Does this mean we're simply uncreative individuals who are unable to bring two *disparate* ideas

together, unable to fully engage in the "mysterious, serendipi-
tous process of creativity"? Is it an indication that we lack the
genes that only very few have who can make large jumps?

Remember that the landscape of your space of possibility is
determined by your assumptions. The diagram is constructed so
as to make certain assumptions, and thus patterns of activity in
your brain (attractor states), more likely than others. In a similar
way, look at the image of the cube below (we'll come back to the
matches and the candle in a bit). You'll see a dark surface on the
top of the cube-like structure, with a light surface below and a
gradient between them. Notice that—without you necessarily
being consciously aware of it—the arrangement of the surfaces
and gradient are consistent with illumination coming from
above. Interestingly, because we evolved on an earth where light
comes from above, this aspect of nature has been incorporated
into assumptions wired into our brain. Consequently, I've
created this image (in a study on the Cornsweet Illusion with

Dale Purves), and in particular the gradient between the surfaces, in such a way that it confirms this assumption in your brain and activates the particular rail tracks that result in you seeing the two surfaces as being differently lit.

Yet, the two surfaces *are physically identical.* They're the same color.

If you don't believe me, look at the image below, where two central areas of the surfaces have been printed in isolation. You can also turn the book upside down, which makes the stimulus inconsistent with light from above. When this happens, the strength of the illusion reduces because the stimulus is now made less consistent with your assumption of light from above.

What is true for the cube is also true for the candle and matchbox creativity challenge. In the same way that you can't see the surfaces in the cube for their underlying physical reality, neither can you see the actual solution to the candle challenge . . . because you're seeing the most likely perception given the space of

possibility of your assumptions. But, as you did with the spinning diamond, changing your assumptions can change the next possible perception. This puzzle is difficult not because the solution is itself difficult. Rather, it's difficult because you are blind to the assumptions that the stimulus was designed to activate. If you knew what those assumptions were, the solution would come immediately, since the solution would be the most obvious answer given the new set of assumptions required to solve the puzzle.

So let's reconsider whether we *are* uncreative individuals, or instead actually just people who are only naïve to our assumptions . . . assumptions that are constraining what is possible for us to see and think and imagine. In exploring the latter possibility, take a moment and try to reveal your assumptions about each element of the picture to yourself.

What are your assumptions about the thumbtacks? What are your assumptions about the matchbox? What are your assumptions about the candle itself? Articulate these assumptions clearly; even write them down if you like. Why do you think the thumbtacks serve one purpose and not another? As an extension of this question, ask another . . . and another . . . and in doing so, you are beginning to see yourself see in real time. Asking why in this manner is useful not only for challenging assumptions but for finding them as well. This approach allows you to hold multiple, even contradictory realities in your mind at the same time, and evaluate them. Are the possibilities different now?

In the first image the thumbtacks are inside the box. As such, your brain makes a perception of least resistance and assumes that the box is a container, and that containers contain things. It couldn't be anything else, until you consider the possibility that it in fact *could* be something else. Thus, when you view the second image where the

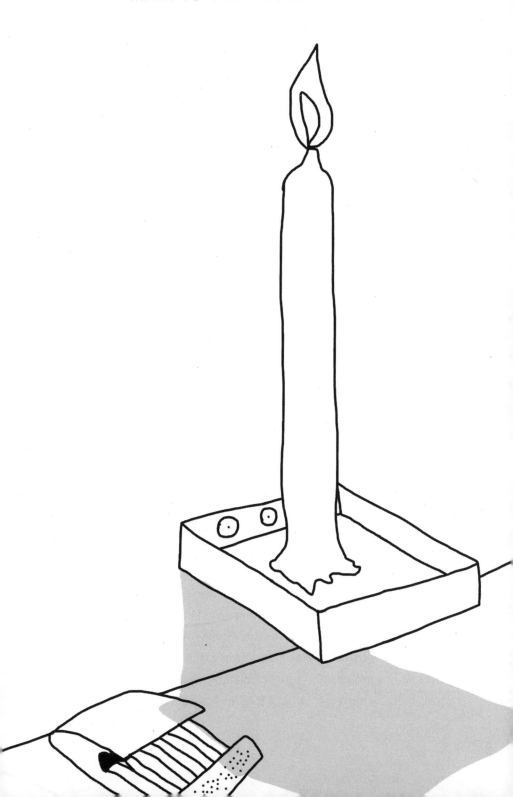

tacks are placed outside the box, this stimulus frees your brain from the first assumption ("container"), and the idea of a "shelf" becomes possible. Of course, you could have taken the tacks out of the box yourself in your mind. But it didn't even occur to you to do so, since even the question itself was constrained by your assumptions. But for those who did change their assumption about the box (consciously or—more likely—unconsciously), the answer would have come in an "aha" moment. But is this *really* an aha moment?

It wasn't! There was nothing particularly miraculously inspirational about it. Rather, like all things creative for the person being creative, it was a small, logical progression to the next possible perception, given good questions that led to new assumptions that rearranged the space of possibility, which **happened** to have a big effect. What was blocking our creativity with the candle was not a genetic predisposition to linking disparate ideas, but rather our species' genetic predisposition to be blind to the causes of our own perceptual behaviors. This is one of the most important obstacles to seeing differently and deviating from uninspired, conventional perceptions: **our assumptions are as blind to us as we are to them.** This bias-blindness is the foundation of what I referred to earlier as the Physics of No. The person saying no doesn't necessarily know why, but only that they always have done so in that circumstance previously and so behave *as if* their perception is an immutable law of nature. This in turn makes it impossible to question, until we gain sight (awareness of our assumptions). But why do we continue to be blind to so many crucial assumptions even after learning that we ARE our assumptions? And how do we develop a method for constantly uncovering them, and thus enable our dormant creativity? How do we cure this kind of conceptual blindness that has led to so much destruction, from religious zealotry to everyday bigotry?

One of the main reasons we are so often blind to our assumptions is that the brain tends to perceive itself as stable and unchanging. Yet, to quote my partner quoting one of our favorite poets, Walt Whitman: "I contain multitudes." For poetry, this is a lovely metaphor that captures a nuance of the human spirit. For neuroscience, it's a proven fact that captures the truth of perception. As the cognitive scientist Bruce Hood wrote, "Who we are is a story of our self—a constructed narrative that our brain creates."[65] In this sense, we are like one of my earlier lightness illusions ourselves, a color that changes shade depending on what is around it. Our assumptions—and thus our personalities—differ depending on the day, the place, and the company we keep. Yet we rarely approach life with this in mind. It makes us nervous.

Evidence of this is that companies are spending millions if not billions of dollars on the pretext that people are stable and unchanging. There is a huge push in twenty-first-century business to define who someone is. What companies are really asking is, "What are their assumptions?" This started in a big way with Google and Facebook, as they figured out the "value" of information. But Google and Facebook don't want information per se, since, as we know, raw data is as meaningless as stimuli falling on the retina. What they want to know is *why* you search for what you search for. The semantics of keywords are a proxy for this. Much of Silicon Valley is now centered on funding enterprises that are able to bring in unique data about human assumptions. But this trend started decades ago with so-called targeted marketing. In the end, it all comes down to selling to someone's biases. But you can't do this well unless you know what those biases are . . . to know another person's why. So what many do instead, quite naturally, is approach consumers as if we were non-contextual, unchanging personalities . . . that is, as if we formed a composite average. It's the part of the road that is illuminated and thus measurable. But so-called measurability is inherently limiting; moreover, it is fundamentally out of synch with the realities of life lived with others.

What would you do, for instance, if you were in a romantic relationship with someone who treated you as if you had the average assumptions of humankind? What's more, imagine that this partner gave you the average amount of affection, spent the average amount of time with you, had average sex with you the average number of times per week in the average kind of way, shared the average amount of feelings, and hit the milestones of getting married and having kids at the average age (but hopefully not the average number of children, lest you be left with a fraction). Surprisingly, because humans are all variations on a theme, it's not a bad strategy. Without bruising our egos too much, it's likely that they'd do OK with this strategy . . . initially, that is, since as humans we all exhibit basic shared assumptions. Most relationships start this way . . . indeed most engagements with the unknown do, quite practically, since this is all one has to go on in the beginning.

However, is it a good strategy longer term? Definitely not. Engaging with people in the world as if they were an unchanging, measurable "average" can have disastrous effects.

In 2014, Victoria's Secret launched "The Perfect Body Campaign" in the UK. As usual, the advertisement photos for the new line of undergarments featured curvy, paper-thin supermodels that made most potential customers feel inferior, even as it was trying to motivate them to buy. Instead of responding to the campaign with frenzied, "aspirational" buying, as Victoria's Secret had hoped, indignant women launched an outraged counter-campaign. A Change.org petition gained viral momentum online with this clear request: "We would like Victoria's

Secret to apologise and take responsibility for the unhealthy and damaging message that their 'Perfect Body' campaign sends out about women's bodies and how they should be judged. We would like Victoria's Secret to change the wording on their advertisements for their bra range Body, to something that does not promote unhealthy and unrealistic standards of beauty, as well as pledge to not use such harmful marketing in the future."

Victoria's Secret had failed to market to its potential customers' real identities, but even more important, the company hadn't marketed to their potential customers' ideas of beauty, which in reality encompassed many different shapes and idiosyncrasies. Instead, the company had spoken to a single idea of beauty (which often gears toward an idealized "average" of beauty) that had nothing to do with the diversity of people's personal lives and the diversity of how we collectively conceive of beauty. Their mistake was less in presenting the so-called unattainable than in thinking they knew the assumptions that created the spaces of possibilities of their customers . . . namely, that women all aspire to one "perfect body," which they don't. It's simply not possible, so people want to feel reflected in their own uniqueness by a brand, not excluded by it. In the end, Victoria's Secret discarded past assumptions about what would be useful in a campaign for driving sales . . . or at least sort of. They kept the same photo for the ad campaign, but gave it a new tag: "A Body For Every Body."

Victoria's Secret behaved as if they didn't know why they did what they did; and alienated the very people they wanted to connect with, which is quite simply bad business. Nor—it seems—did the company understand their consumers' spaces of possibility, or that those spaces are *contextual* for each of us, though they quite likely did focus groups and other types of research to pretest their campaign, which clearly failed. If businesses want to understand their customers better, they should focus less on common market research techniques like focus groups and questionnaires. These are only reliable in stable environments with consumers that, well, aren't human. As noted above, humans are not only unstable but also naïve to themselves. Hence, research by social psychologists has shown that the answers people give tend to *not be true* to a surprising degree. The evidence suggests that what respondents are doing is answering in a way that reflects who they would like to be, not who they actually are. Also, sometimes they're not paying attention, or even take glee in responding untruthfully. This is why we must conduct research *"in our natural habitat,"* where real decisions are made for reasons of consequences with feedback that can be either trivial or fundamentally salient. As such, it's much more reliable to study people's *behavior* than their introspective interpretation of it. A useful perceptual principle: If you want to understand humans or a situation generated by humans, you need to know their assumptions. But don't ask them! And if you want to understand yourself, sometimes the best answer exists in *another person.* Studies have shown that other people can be better at predicting our behavior—at usefully predicting us—than we ourselves are. That said, we do have an important built-in method for seeing ourselves more clearly . . . for not treating ourselves like "an average" and for empowering ourselves with our deviance and questioning.

What is this method? Emotions. The physiology of feelings often tells us everything we need to know. This is important, since – among other things – our emotions are indicators, or proxies, that reveal one's assumptions to oneself (as might have happened to some when learning about the positive data of psychedelics on the lives of individuals). When we go into any

given situation, we have expectations. Your brain rewards your ability to predict the future, including the behaviors of others, by releasing good-feeling chemicals in different regions of your cortex. When you fail to predict correctly, however, the opposite of tranquility occurs: You feel negative emotions, and your perception is flooded with all the accompanying sensations. Your

assumptions failed you, meaning that a useful predictive perception wasn't within your space of possibility. Thus, the negativity
that you feel upon entering a conflict (with yourself, with another,
or with the world) is simply a reflection of the fact that what is
happening now is not consistent with what you think "should"
be happening; that is, the meanings from the present don't match
past meanings. The interesting thing is that when you are consciously aware of your assumptions going into a situation, even
if your assumptions aren't met, your emotional reaction will be
less severe because you weren't blind to the forces that were
shaping your perception in the first place. My past colleagues at
University College London have discovered an equation that
roughly predicts whether someone will feel happy or not in a
given situation.[66] The basic key to their finding is whether the
event matches expectation. What is more, the expectation of an
event matching your expectation alters your level of happiness
in anticipation of the event itself . . . that is, until you actually
experience it. It's in the anticipation of an event that happiness
itself often lives. This is because dopamine—a neurotransmitter
in the brain correlated with positive feelings, among many other
things—spikes in anticipation of a positive event, and then
decreases during the event itself.

To make the point . . . to put the challenge of creativity into a
more immediate biological context . . . consider seasickness. We
sometimes get nauseous on boats, especially when we're below
deck. One reason is because there is a conflict between two of
our senses. As our eyes move in register with the boat around us,
they effectively tell our brain: "We are standing still." However,
because our inner ears are receiving signals that indicate movement, they tell our brain: "Nope, we're definitely moving." So
there is a conflict between the vestibular and ocular systems in
your brain. Your brain has entered uncertainty. A key response to

internal biological conflict is nausea, which is the brain's way of telling the body, "Get out of here!" To remove the conflicting uncertainty, you can either lie down and close your eyes (thus removing the battle of contradictory information), or you can go above deck and watch the horizon, so that the sensory input from your eyes and inner ears falls back into a complementary certainty. Now the world makes perceptual "sense" again, and your body calms. When your body isn't calm, however, the stress induced by uncertainty drastically narrows one's space of possibility in the pursuit of maximizing efficiency and sharpening physical responses.

While sea sickness isn't strictly an emotion, knowing that this is how emotions work in general is extremely useful. First, it gives you a practical task when something upsets you, big or small . . . to ask yourself what was the assumption that wasn't fulfilled. This postmortem of a situation, be it professional or personal, is so valuable because it forces you to look for the unseen assumption that guided your behavior. Once you see that assumption, you have the potential for choice.

This means that to truly uncover your assumptions and re-place if not expand them with new ones, you must constantly step into an emotionally challenging place and experience difference . . . *actively*! By challenge, what we really mean when we say this is that an experience or environment doesn't match our assumptions (expectations). But here this emotional challenge is in fact a good thing, since actively seeking contrast is the engine that drives change (and the brain). Thus, *diversity of experience* is a transformative force for the brain because new people and environments not only reveal our assumptions to ourselves, they also expand our space of possibility with new assumptions. A particularly useful way to experience difference of contrast is to travel.

In recent years, there has emerged a growing body of research about the effects of openness in travel on creativity and the brain. In collaboration with several colleagues, Adam Galinsky, the researcher whose work with enclothed cognition we looked at in Chapter 5, is pioneering studies in this new area. In a study published in 2010, Galinsky, William Maddux, and Hajo Adam found that having lived abroad was strongly linked to heightened creative intelligence. For the experiment, which was conducted in France with French college students, they primed the participants, who had all previously lived in other countries— that is, had them do an activity that triggered a certain state of mind—in two different ways and on two different occasions, then asked them to carry out the same task. In the first prime, participants had to write about an experience in which they "learned something new from a *different* culture." In the second they had to write about a time they "learned something new about *their own* culture." In both instances, after the priming task participants had to do a word-completion test similar to the letter-string exercise you did earlier. What was the result? The participants were significantly more creative when primed with their multicultural experiences abroad. In other words, their space of possibility took on a higher dimension (effectively becoming more complex; more on this later in the book) when drawing on the perceptual history they had gained in another country. Living abroad had already uncovered many assumptions that structured their past perception, so they were less tied to their previous biases. In addition, they were more capable of seeing calcified assumptions as such. How we act once we unblind ourselves from our biases, however, depends on each individual's willingness to deviate and—of fundamental importance—the ecology that they find themselves in and/ or create.

I find there are often two general responses to living overseas. One response is to become a more extreme version of oneself (becoming the prototypical "expat," who resists the "deviations" of a foreign culture), where one's original assumptions become more stable *because* they are in contrast to differing assumptions of another culture. There is considerable psychological evidence that shows when people are placed in a new, uncertain environment they can often become more extreme versions of themselves.[67] A true irony (which, of course, the brain is full of) is that when presenting people with more and more rational reasons that their assumptions are wrong, they will hold increasingly fast to their argument, which quickly becomes more faith than knowledge. We witness this in the current experience of the climate change "debate." Note that while the BBC and other reputable news agencies must offer equal time to any two sides of the argument, this doesn't mean that both sides have equal support for their view.

The other, more enlightened common response to life abroad is to embrace the (*thoughtfully positive*, constructive) assumptions of another culture. The consequence is a space of possibility of higher dimensionality, and thus by definition one of a greater degree of freedom (i.e., diverse in both the mathematical and metaphorical senses). I've had the immense pleasure of experiencing difference in others with tremendous positive effect. One of my graduate students when I was at University College London was from Israel, where a number of people I've had the honor of being friends with hailed from (along with Scotland, England, Chile, the United States, and others). Despite spending her whole life up to the age of 22 in Israel, she had never actually met a Palestinian in any deep way. This was hard to believe for someone like me, an outsider whose only sources of information about another location are superficial, but there you go. When she moved to London she—of course—needed accommodation. It just so happened that the flat in which she found a room was located in an area of London with a particularly high density of Palestinians. By coming to London, she was able to experience the assumptions of the Palestinian community that had been in fact omnipresent in Israel. In doing so, she redefined what "home" was when she returned, seeing Israel and its people differently, and maybe even more important, seeing herself differently. Thus, multiculturalism is a positive attribute, and indeed a necessary one for any metropolis that wants to encourage innovation, given a certain way of being (as it isn't just diversity that matters, but also the ecology of innovation that diversity inhabits . . . more on this later). Exposure to a diversity of ways of being has been shown to increase not only empathy, but creativity itself.[68]

In another study that Adam Galinsky and his colleagues conducted, titled "Fashion with a Foreign Flair," they came away with even more resounding findings, as they left the lab to examine creativity as it is linked to success in the professional world of fashion. They studied 21 seasons (11 years) of collections of world-renowned clothing design houses and concluded that "the foreign professional experiences of creative directors predicted the creativity ratings of their collections."[69] That's right: if the leaders of the company had not just lived but also worked abroad, they were more creative, and this trickled down to the culture of the entire company in the same way that a host determines the demeanor of a party, as does the leader of a company who takes on the assumptions of leadership. It could be argued that stocking the workplace with at least some people who have lived abroad makes it in principle stronger as a whole. The foreign experiences of leaders inside the company will have the most influence, but at any level, the more elastic the assumptions, the better . . . and not only for cerebral tasks but for interpersonal ones, too. A 2013 longitudinal study published in the *Journal of Personality and Social Psychology* showed that when we "hit the road," people tend to become more agreeable and open, and less neurotic. Naturally, being more open increases our chances of espying the assumptions shaping our perceptions. So . . .

Be an alien. Strangers in strange lands bring home new brains. This makes it easier to see assumptions that need questioning, and then actively question them.

Yet there are limitations to the benefits of such globetrotting lifestyles; or rather, there is a sweet spot professionals residing abroad should strive toward. Galinsky's fashion study found that too much international movement, or immersion in cultural environments too drastically dissimilar from one's own, could

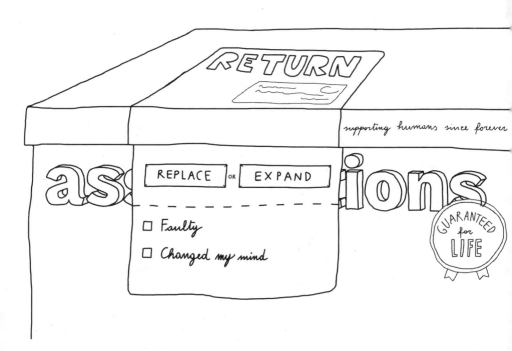

actually inhibit creativity. This very likely has to do with challenging situations unloosing stress hormones in the brain, which
overrides the free-flowing electrical patterns of perception associated with creativity in favor of more urgent, fight-or-flight-style
responses. Recall Marian Diamond's work with rats raised in
differently enriched environments, where the more enriched the
environment the more connected the brains of the rats became.
But when the environment became over-enriched, the brains of
the rats actually *decreased* in complexity. Deciding on what is "too
much" is of course personal. The lesson is that living and working abroad can be very good, but it's advantageous to do so in
places where you're not completely starting from scratch culturally, linguistically, and economically.

Not all of us, however, have the opportunity to travel abroad
for the pleasure of revealing assumptions (among others), much
less to live and work abroad. Does this mean that we'll be out-
witted by our competitors who have? No. To reveal assumptions
that will in turn open up the possibility of a new future past, our
travel doesn't have to be to the other side of the world. If you are
willing to ask why, it can be just as generative for your brain and
perception when it is within your own country, whether in
a neighboring state, neighboring city, or even a neighboring
neighborhood. What is crucial is that you are challenged with
contrasting situations and ecologies—and that emotions are
provoked—since these will push you into unfamiliar contexts of
trial and error that will then become recorded in your neural
history of responses. You can also travel in your mind.

As we already know, imagined experiences encode themselves
in our brains in nearly the same way as lived ones, so engaged
delusion can offer a substitute for travel, which might explain the
popular genre of travel writing (and indeed reading in general,
since books take us into unfamiliar worlds). As surrogates or

stand-ins for us, the travel writer's process of revealing his or her own assumptions becomes our own, and perhaps there is no one better at describing the often chaotic and difficult, yet often life-affirming experience of travel than the American-born writer Paul Theroux.

An inveterate adventurer who has ridden the Trans-Siberian Railway, trekked through Afghanistan, and visited Albania during its civil strife, the entertainingly grouchy Theroux also writes movingly about not just travel but why we travel: "The wish to travel seems to me characteristically human: the desire to move, to satisfy your curiosity or ease your fears, to change the circumstances of your life, to be a stranger, to make a friend, to experience an exotic landscape, to risk the unknown."[70] Also he writes, "Delay and dirt are the realities of the most rewarding travel."[71]

So get dirty.

While reading is powerful, we do reveal our assumptions and make new ones most starkly, and by extension benefit most perceptually, *by physically engaging* with the world . . . by reaching out! This is how the *It's real-life trial* brain most effectively and *lastingly* makes *and error, and* meaning and also re-means past meanings. It's *it's not only how* real-life trial and error, and it's not only how *we uncover our* we uncover our biases. It's how we actually *biases. It's how* *make* new ones. *we actually*
make new ones.

Meet Destin Sandlin. He is a missile engineer from Alabama who is married with two kids. He has infectious energy, a charismatic southern accent, and a round-cheeked smile he seems rarely capable of turning off. He is also a person who has done the following things: shot an AK-47 underwater, studied the physics of a "slow-motion flipping cat," had a hummingbird eat from a feeder in his mouth, milked the most venomous fish in the world, analyzed the lip movements of people beatboxing, eaten goat brain, and recorded the work of tattoo artists at 32,000 frames per second. Come to think of it, pretty much all of these things he filmed at an extremely high number of frames per second, and they are quite astonishing to see. Why has Sandlin done all of this?

The superficial reason is that he has a very popular YouTube science show called *Smarter Every Day*, with well over 100 episodes to its name. The real, deeper reason is that he is a person who by nature and by choice seeks out wildly diverse life experiences that embody questions, even if surely he's too self-effacing to put it this way. He frames it as curiosity about how things work, why things happen the way they do, and how science can explain the most enigmatic-seeming things, like why the legendarily steel-stomached Houdini died from a punch to the stomach, and whether toilets flush in different directions in the northern and southern hemispheres (they do). All of this is just an elaborate package for what Sandlin is really doing: adding new experiences to his perceptual past that constantly reveal his assumptions about the world around him. He is seeing a version of life that is oriented not toward average experiences, but toward his personal (thoughtful and thought-provoking) deviance. In so doing, he changes what is possible in his brain in the future.

In one episode, titled "The Backwards Brain Cycle," a welder at the company where Sandlin works specially designed a bike so that when he turned the handlebars left the bike turned right, and vice versa. Sandlin's first assumption was immediately revealed: He thought he wouldn't have any trouble riding it. He would intellectually incorporate the reversal into his motor-processing neural circuitry lickety-split and pedal off. Sorry, not quite. He challenged this assumption, embodying his questioning as a *lived experience* in the world, something that physically happened to him . . . and it took him *eight months* to learn to steer the bike, riding it every day for five minutes. (It took his young son only two weeks, thanks to his young brain's neural plasticity, which results in an ability to incorporate feedback (read: experience; trial and error) faster than adults.) What I particularly like about Sandlin's work is that he too believes in *living* (and thus becoming) the thing you're talking about. Thus Sandlin has taken the bike with him for talks in places as far-flung as Australia, and nearly every time someone cockily volunteers to ride the bike onstage . . . then falls. In this way, Sandlin's way of being and space of possibility shifts that of others. By sharing his lived questions that uncover assumptions, and making people physically participate, he changes their brains and perception. He gives them potentially new reflex responses.

Sandlin's experiment with the Backwards Brain Bicycle challenged a bias he had trained into the "physics" of his neuromuscular cells to the point of invisibility: steering left directs me left, while steering right directs me right. This was such a deeply entrenched assumption—an attractor state if there ever was one, a rolling wave whose momentum seemed impossible to stop—that his conscious awareness could only unseat it after considerable and time-consuming effort. It was a "body assumption," but it all came from his neuroelectrical network, since there is no universal law that says how bikes must steer (no matter how intuitive). He uncovered the reflex arc that past experiences had given him, and the behavioral limitations this imposed on him. More than this, he was also able to embody new assumptions: In a strangely exciting moment at the end of the "Brain Bicycle" episode, Sandlin goes back to riding a "normal" bike for the first time in eight months. Or rather, he *tries* to go back.

Sandlin carried out this stunt in Amsterdam, one of the most bicycle-friendly cities in the world, and bystanders were clearly puzzled by this adult who appeared to be riding a bike for the first time. What was really happening, of course, was that his brain was struggling to reroute its currents from the newly attained neuroelectric attractor state for how to ride a bike back into its old more conventional reflex arcs. But then he did it. "It clicked!" Sandlin said, suddenly able to ride again, though still with quite a bit of wobble. "It clicked!" . . . a physical manifestation of a behavioral self-criticality when one small movement seemingly led to a big effect in the moment. He was suddenly able to institute useful new assumptions in his brain only because he had rediscovered the old one and its meaning. I wonder whether with more time and work the new assumption could coexist with what at first seemed a set of contradictory assumptions, and he could become "ambidextrous" at riding the two types of bikes. Most likely he could, instituting more dimensions (and coexisting, contradictory-seeming assumptions) in his perceptual network.

Sandlin never could have trained his brain to go from A to B . . . namely, to be fluent in these "dual" perceptions . . . if he hadn't forced himself to physically engage with the world, which allowed him to see the unseen forces shaping his perceptions. From this he was able to challenge the assumptions . . . as we learned to do in the previous chapter by examining how great ideas are "made" . . . and then embody the new assumptions through the feedback inherent in empiricism and trial and error. In doing so, the novel experience he created for himself opened up invigorating new possibilities (and the requisite connections) for him and his brain, as well as for anyone else willing to risk taking a tumble.

Key to Sandlin's discovery of assumptions and both the

creation of new ones and thus the expansion of his space of possibility was his bike. A novel technology is often key to the process of deviation. By "technology," I don't mean the latest app or device (though I'm not necessarily excluding them either). Most technologies make what we can already do easier, faster, or more efficient, which of course is useful. However, the best technologies—I argue—are those that impact our awareness of previously unseen assumptions, as well as change and expand them, with a resulting change in our individual and collective spaces of possibility. Hence, the greatest innovations tend to be those that show us new realities, like the microscope, the telescope, the MRI, the sail, theorems, ideas, and questions. The best technologies make the invisible visible.

> *The best technologies make the invisible* **visible.**

Such technologies open up new understanding, transforming the ideas that we have about our world and ourselves. They not only challenge what we assume to be true already, they also offer the *opportunity* for a new, larger, more complex set of assumptions. They push us further from being the center of the universe toward the more interesting and interested perspective of an outsider.

The point about perception here is that if you want to get from A to B in your life, whether you're transitioning personally or professionally, the first challenge is to accept that *everything* you do is a reflex grounded in your assumptions. So we need humility. Though change never happens without it, by itself that is never enough. Once we come to accept that all we see and do is grounded in our assumptions, we are nonetheless usually blind to the reasons why we do what we do. The next challenge to deviation, then, is to discover what your assumptions are. This usually involves other people who are foreign to you, hence the power of *the diversity of groups*. The next step is to complexify your assumptions—and thus redefine normality—by *actively engaging* in the contrasting nature of the world. That's what Destin did, as did the people who wore the feelSpace magnetic belt and acquired heightened powers of navigation.

Another key for seeing differently is not to move through the world comfortably. Whether literally or metaphorically, in one's body or in one's mind, we need to get dirty, to get lost, to get swallowed by the experience. This could sound clichéd, but it's nonetheless true . . . and necessary to reiterate loudly, given the speed of the sprint in which much of the Western world is running toward health and safety. (We're rushing so fast toward mitigating against short-term risk that to stand still in our society is to become relatively risky!) Don't be a tourist of your own life, taking your assumptions with you wherever you go. Leave them in the lift at JFK or at Terminal 5 at Heathrow. And when you get to wherever you're going, buy groceries, ask for directions in the local language, navigate an unknown transit system, try to remember how to get back to your hotel rather than referring constantly to your Google map. And through it all, listen to your emotions so you know whether you've traveled far enough. Only in this way will you be able to discover the mishaps and misfits that all-inclusive luxury vacations rob you of. Only then will you discover the invisible in yourself by *assuming you might be wrong about your "knowledge" of things*. Seek new, more generalizable assumptions through real-world engagement in concert with your delusional powers, and in doing so, you will alter the probabilities of your *future* reflexive responses by increasing the chances of beating the kurtotic biases that past experiences have given you. This is how you institute new and better assumptions and "travel" to new perceptions. In short, don't shift . . . expand!

But once you have unblinded your assumptions, experimenting with new ones is not an easy process, and our very own evolution often steers us away from it. Our brains want to avoid it, even if the results will be generative for us. This brings us to the second biggest challenge to creativity: we are afraid of the dark.

CHAPTER 9

Celebrate Doubt

Darkness. Few things frighten us more. The fear it creates is a constant in our existence: The living darkness of our bedrooms after our parents turn out the lights. The pregnant darkness beyond the glow of the bonfire as we listen to "spooky" stories. The ancient darkness of the forest as we walk past deep shadows between trees. The shivering darkness of our own home when we step inside wondering if we're alone.

Darkness is a fundamental, existential fear because it contains all the fears that we carry with us in our brains—fears both real and imagined, engendered from living life and from the life lived in stories, from culture, from fairytales. The more we are in the dark, the more there is to fear in the dark: hostile animals, bone-crunching falls, and sharp objects that will draw blood; muggers, rapists, and murderers; and the imagined creations that aren't even real but frighten us just as much: evil spirits, mythical beasts, and the flesh-eating undead. Darkness is the embodiment of the unknown, and this is what scares us more than anything else: knowing that we don't know what dwells in the space just beyond. Not knowing if we are safe or unsafe, if we will feel pain or pleasure, if we will live or die. Our heart thuds. Our eyes dart. Our adrenaline courses. The unknown "haunts" humans. To understand why this is, we must go back in time and look at how this fear made us who we are, and how it helped us survive. This is our evolutionary past, and it explains why creativity often eludes us even when we know how simple the process is. It also explains why asking *why* and unseating bad assumptions can be so difficult.

Imagine our planet as it was nearly two million years ago, in particular the deadly, unpredictable expanse of East Africa, from where we all originate. Great tectonic shifts have caused the landscape of a once flat, forested region to radically transform into a dry, mountainous topography of peaks and valleys, lake basins and plateaus. It is a rugged world of scarce and scattered resources (food, water, tool-making materials) that made our genetic forebears come out of the trees, and in doing so become bipedal—and thus they have not disappeared, as other related species have before them. There were still various types of hominids competing and evolving on this harsh landscape, braving a changeful climate full of other dangerous animals, among them hippopotami bigger than our modern-day ones, wild pigs nearly as large, and vice-jawed hyenas. But only one species of the hominids would survive: *humans*.

This was the environment we evolved in—that our brain and our perception evolved in. It was Earth long, *long* before law and order. It was a highly unstable place in which we had very limited knowledge of sheltering, feeding, and healing ourselves. Humans weren't yet the "masters of the planet" we are today (though true "master" living systems like bacteria and cockroaches will be here long after we are gone). Simple illness killed, as there were no medicines or even the imaginable possibility of them. The world was a hostile and erratic place—the epitome of

uncertainty, with the future cloaked in "darkness." In such a context, not being able to predict was a very bad idea, and predicting
was a very good idea. If you failed to predict where a close
water source might be, if you couldn't predict which plants to eat
and which not to, if you couldn't predict that the shadow "over
there" was something that could actually eat
you in time . . . *it was too late*. Certainty meant
life. Uncertainty meant death. To not "know"
was to die.

*To not "know"
was to die.*

Throughout evolution, it has been harder to stay alive than to
die. As a matter of fact, there are simply more ways to die than
to not die. When you are sitting in your community, sheltered
and protected, where everything is momentarily predictable, the
last thing you want to do is say, "Hmmmm, I wonder what is on
the other side of that hill?" Bad idea! It seems obvious that the
probability of dying just suddenly increased considerably. Yet
what might be true for the individual is not necessarily true for
the group or species. It is because of that "mad" individual that
the group has a better chance to survive *in an environment that is
dynamic*—by learning what dangers or benefits are on the other
side of that hill, and perhaps discovering new spaces of possibility that the group hadn't known existed. Thank God we have
such *seemingly* pathological people out there . . . the deviators
(i.e., "seemingly" to the normal . . . who are the average, and, by
definition, non-deviating).

For fish it's the same. In schools of fish, the ones that stray
away and find food are usually the same ones that are first to get
eaten. In the end they benefit the whole school, even if they are
sacrificed in the process. A diverse ensemble—the full orchestra—
is essential in a world like ours that is always changing! Indeed, it
is common knowledge that diversity is essential for systems to
evolve. Our research on evolving artificial life systems has shown

that diverse populations are more likely to find the optimal solution than a less diverse population. What is more, they were more likely to exhibit more "contextual" behavior (i.e., conditional behavior) when they evolved in **uncertain environments** (where there was a one-to-many relationship between stimulus and reward) relative to **unambiguous environments** (where there was a one-to-one relationship between a stimulus and reward). **Their neural processes were also more complex.** For instance, they evolved multiple receptor types in their artificial "retinae," which is a prerequisite for color vision, when the light stimuli were ambiguous. These findings are consistent with the view that contextual behavior and processing are born from uncertainty.

Overcoming uncertainty and predicting usefully from seemingly useless data is arguably *the* fundamental task that the human brain, as well as all other brains, evolved to solve . . . hence why existing in uncertainty is exactly what our brains evolved to avoid. Living systems in general hate uncertainty. This is why a fear of the dark appears not just in humans but across all simian species, as it makes us tremendously vulnerable.[72] It's not simply the literal darkness that we try to avoid, but also the fear we feel when stepping into the uncertainty of both literal *and* metaphorical darkness. In contrast (yet deriving from the same perceptual principle), rats fear not the dark, but the light. As nocturnal animals, they feel more certain and less threatened amid darkness, since in darkness they have shelter from being seen.[73] For them it is the light, and not the dark, that is uncertain. How amazing! So all living systems (individually and collectively) evolve and learn their own particular "darkness," as well as a deep, active resistance to that particular uncertainty, just as each living system evolves physiological responses for dealing with unavoidable uncertainty. These responses are visceral and physical, and point up the way brain and body are very much one

when it comes to our perceptions and their ability to limit otherwise natural creativity.

Our brain has evolved two general strategies when presented
with the fear of uncertainty. One of them is anger. Have you
ever seen a furious traveler yelling at a hapless gate agent in an
airport? They are often doing so because they're dealing with
the inherent uncertainty of travel; they don't exhibit much innovative thinking or compassion in dealing with the situation,
because their brains have entered a state that prevents them
from doing so. Anger's effects on our perception are in a sense a
cure for fear arising from uncertainty.[74] Anger tends to make us
feel wholly justified in our perception of what is true, which is a
very powerful perceptual delusion of certainty. This process is

supported by the drastic physiological changes in your body that accompany anger. Your muscles tense, and neurotransmitter chemicals known as catecholamines are released inside your brain and increase the perception of energy, which provokes the common anger-fueled desire to take immediate protective action. Your heart rate accelerates, your blood pressure rises, and your respiratory rate also increases. Your attention locks onto the target of your anger and you can pay attention to little else. A person who has been angered will often continue to ruminate long after the stimulus has passed, re-creating the stimulus-response sequence in their imagination and further *reinforcing* the causal neural connections (remember back to the power of imagery on activity in the brain). Then additional brain neurotransmitters and hormones (among them adrenaline and noradrenaline) are released that trigger a lasting warlike state of arousal. And what is interesting and ironic is that the more creative and intelligent the individual, the more difficult it can be to dissuade them from their anger response, since they are better at finding seemingly meaningful connections in what are in fact non-causal relationships, as well as creating internally consistent arguments that support their erroneous position, thus protecting them from the uncertainty of ignorance. But although this "natural" response can have tremendous advantages in certain life-and-death contexts, is it really the best response in all contexts?

The brain's evolutionarily foundational drive to increase certainty gives us a new framework through which to view the Lab of Misfits study on power that we discussed earlier. In that experiment, we primed the subjects into states of lower and higher power (and also a control neutral state), which affected their perception. For example, the group in the lower state saw my color illusions more strongly than the group in a higher state of power; they were being more "promiscuous" with contextual influences, in a sense more gullible. But this gullibility in perception was actually an attempt to *increase certainty and power*. The low-power subjects' sharpened vigilance in perceiving the illusions was a mediator for the aversive physiological state of uncertainty. Younger people also perceived the illusions more strongly because children and the young are in a kind of perpetual low-power state and are very aware of this lack of certainty in the outcomes they effect (though parents who have had a kid melt down uncontrollably in a public place would probably aver their own powerlessness; tantrums are a form of control). Outside the "lab," not understanding how the search for certainty drives us—and constricts our space of possibility—can be tragic, though it also explains self-destructive behaviors, even though they may feel "logical." Just ask Donna Ferrato.

Ferrato is a woman with the penetrating gaze of someone who has seen a lot . . . because she has. As the fearless photographer whose photos of domestic violence forced the issue into the public spotlight in the 1980s, she has seen—and captured—the dark side of human nature up close. Her seminal book, *Living with the Enemy*, is a devastating portrait of women under siege in their most intimate relationships. Her online multimedia project, *I AM Unbeatable*, explores the journey toward a new life, observing women who escape their abusive environments. This process, however, is not simple, and Ferrato has watched women take years before they finally chose the uncertainty of leaving over the certainty of continued violence. This tendency tells us about the strange extremes toward which not understanding our own perception can blindly push us. It also explains the famous proverb, *Better the devil you know than the devil you don't*.

The drive to mitigate uncertainty underpins much of what we do, feel, see, experience, choose, and love. This is true to the extent that we even prefer to experience pain over uncertainty. A recent study by scientists at UCL shows that the greater the level of uncertainty, the more stress participants experienced. According to the lead author of the study, Archy de Berker, "It turns out that it's much worse not knowing you are going to get a shock than knowing you definitely will or won't." In short, not knowing is more stressing to the system than having certain knowledge of something bad. Hence, controlling the level of another person's sense of uncertainty can be used as a means of control, whether by governments or even within interpersonal relationships.

This is perhaps why sexual "dis-disclosure" (which is only one form of transparency) can enhance a sense of closeness in a relationship (and perceived relationship-satisfaction more generally). And why companies that are in the business of reducing uncertainty can be incredibly profitable. As the insightful marketing guru Rory Sutherland pointed out to me, the success of Uber is less about disrupting the taxi industry (which most ascribe it to be) and more in line with our own research on uncertainty, in this case our need for certainty about where the taxi is and when it will arrive. With this perspective in mind, look around you and you'll find yourself noticing similar examples. As Rory pointed out, stress in bus shelters in London was dramatically reduced when Transport for London installed LED displays telling people when the next bus would be arriving. Even if the wait was considerable, knowing that it was considerable made the experience more emotionally tolerable. Similarly, the designers of Terminal 5 at Heathrow retrofitted an "UP" button in the elevator from the Heathrow Express platform, even though there is only one direction and only one floor to which the lift could go! Without it, people would enter the lift and find no button to push, and then panic. What's beautiful about this design is that the plastic button is not actually connected to anything but a light. So, when the future travelers touch it, all that happens is that the button lights up (and then the lift does what it would have done anyway). This example also demonstrates the deep inverse relationship between our fear of uncertainty and agency. When you feel you have agency (a sense of control—delusional or otherwise), your sense of uncertainty is also decreased. Designing according to this deep neurobiological need has tremendous effects not only on well-being but on the success of the companies that engage in this practice, and thus provides a fundamental design principle that guides our design thinking when creating experiences for others (within the context of the Lab of Misfits).

It is clear, then, that the desire for certainty shapes our spaces of possibility, our perceptions, and our lives both personally and professionally. Usually, this need saves us. But it also sabotages us. This produces an ongoing tension between what we might think of as our conscious selves and our automatic selves. To overcome our inborn reflex that pushes us to seek certainty (sometimes at any cost), we must lead   *Celebrate doubt!* with our conscious selves and tell ourselves a new story, one that with insistent practice will change our future past and even our physiological responses. We must create internal and external ecologies that . . . celebrate doubt!

This means the biggest obstacle to deviation and seeing differently isn't actually our environment as such, or our" intelligence, or even—ironically—the challenge of reaching the mythical "aha" moment. Rather, it is the nature of human perception itself, and in particular the *perceived* need for knowing. Yet the deep paradox is that the mechanisms of perception are also the process by which we can unlock powerful new perceptions and ways to deviate. Which means that mechanism of generating a perception is the blocker . . . and the *process* of creating perception is the enabler of creativity. Through a neuroscience-based, *intentional* engagement with your perception, you give yourself the opportunity to unveil possibilities in your thoughts and behavior that might otherwise be unavailable. Yes, our number-one aim is to stay alive. But to stay alive for extended periods of time requires more than responding in the moment. It requires adaptation, since the most successful systems in nature are also the most adaptable. What is more, we want to thrive in every sense. To achieve this requires taking risks in perceptions, which requires us to be deviant. To deviate well is hard. It takes commitment, since we're not just up against ourselves.

Humans have come to structure their societies around

institutions that provide certainty: courts, governments, police, and, most sadly, our educational systems (even at the university level), and the processes therein. In politics, the notion of a "U-turn" or "flip-flopping" is always portrayed negatively. Just think for a moment how silly this is. Do we really want our politicians—indeed, anyone—to hold on to what they "assume to be true" in the face of evidence (and sometimes overwhelming evidence) to the contrary, especially knowing that what was "true" yesterday isn't necessarily true today? Or do we want politicians (and leaders more generally) who embody flexible thinking, people who are as pragmatically plastic as the brain that produces their thoughts?

Religion also reduces uncertainty for us, which is a principal reason (among other reasons) why so many billions of brains value so passionately the assumptions that their unquestioning faiths espouse. A 2014 BBC article reported that high rates of atheism appear in countries with marked stability, while places besieged with more disaster-prone natural living conditions usually have "moralising gods."[75] But the flip side of this security is that religions replace your own assumptions with theirs, and take it as an article of literal faith that you will not question them. And, of course, there is that momentum of our attractor states of neural activity inside our brains (just like our cultural homeomemes), the inertial force of past meanings that push us toward reflexive perceptions that are hard to resist.

But embracing uncertainty in order to innovate within it is possible . . . in fact, it is essential. Feeling uncomfortable and other sensations of discomfort are in fact *welcome states*. They are places to dwell and let perception explore, like a crowded foreign market with vendors yelling things at you in an unintelligible language. If you can muster the courage to celebrate doubt enough to say, "Fine, I don't know," suddenly the constricting axes of your space of possibility fall away, freeing you to construct an entirely rebooted space full of new ideas. Approaching conflict with doubt is not just possible but ideal. Now that you know how your perception works, you're able to understand that you might be behaving and thinking in a certain way not because it is an expression of your most creative, intelligent self, but rather out of your physical need for certainty.

So how do we ensure that we engage in creative conflict in practice? Most essential, we must *listen differently* . . . that is, *actively observe*. Instead of listening only to help us better defend ourselves, we must listen with the thought that *conflict is a genuine opportu-nity*. And by conflict, I mean this *gen-erally*, namely a situation that doesn't match one's expec-tation, hope, or desire, such as when facing the conflict of gravity for the first time as children, or a view that differed from our own later in life. The most essential space for learning is in con-flict *when in the right* personal relation-ships, that *ecology* requires *both sides* letting go of the idea of a confrontational conversation . . . "I'm right and here's why" (though there are times when this is an essential position to take, just not nearly as many as we seem to think).

Imagine what governments would look like using this model . . . or interpersonal relationships more generally. Imagine what would be constructed creatively if one engaged with difference with not-knowing rather than knowing . . . entered conflict with questions, and not answers . . . with the intention of understanding how this difference might reshape my perception and my future past. Not only can this approach be an essential way to uncover *your* own assumptions . . . as well as the assumptions of the other, embodying this approach will change your future past, and thus alter your approach to future conflict. While this is a radically open and pacific attitude toward fighting, it's not saintly. You are being productively selfish, since you are looking for the best ideas to create more understanding in your life and relationships from inevitable conflicts that arise.

This is why conflict is in fact as much a part of love as physical affection, though for most people it's not nearly as enjoyable. Unlike hugs, kissing, sex, and other kinds of touch (which also require active listening to be felt meaningfully), the emotional touch of conflict is often what divides instead of unites a couple, since it so often leads to the anger axis described above: bickering arguments and worse. It is an outlet for tensions that arise out of everything from banal matters like daily logistics and housework, to attitudes toward weightier matters like ways of being, fidelity, and fundamental world-views. Yet as most of us know, *all* matters can become weighty during conflict if we don't stay aware of how perception works *in oneself, NOT just in the other person*; the pettiest gripes are capable of unleashing the worst arguments, and the messy axis from fear to anger starts guiding our perception of self and—in particular—of others. This is because conflict often comes down to your own assumptions not being fulfilled, which produces emotions that in turn may not meet the expectations of your partner, producing still more emotions, and so on. In other words, two people in a relationship never have the same past, never have the same brain, and thus never have the same space of possibility filled with the same ideas in the same positions of availability. While too many rarely question the validity of their own expectations or what they reveal about themselves (ideally to themselves), conflicts can find resolution when couples find a (literally) "higher" state of being, loving, and understanding that when they expand the dimensions of their space of possibility to include the assumptions of their partners, they co-create a larger, more complex space wherein both responses can not only coexist, but are no longer in opposition. But this process of learning through conflict requires patience and care and—of course—new assumptions about conflict itself.

John Gottman is a psychologist and professor emeritus at the University of Washington who, along with his wife, Julie Schwartz Gottman (also a psychologist), has revolutionized the study of relationships. By bringing a scientific and not just clinical approach to their work with couples, they have pioneered the gathering of data, as well as a method of meaning making for identifying universal patterns of behavior that produce either happiness or unhappiness in relationships. Much of this work they carried out at their "Love Lab" in Seattle, a studio apartment rigged with state-of-the-art equipment that tracks the physiological responses of couples that volunteer to spend time there. The participants are tasked with simply being themselves as scientists observe and record them, while also taking measures of physical indicators such as heart rate. Over the years, John Gottman has marshaled the data gathered from studying couples interacting to design a model that makes him 91 percent accurate in predicting whether they will divorce or not. Now, *that's* an unsettling amount of certainty.

The Gottmans have identified what they call "The Four Horsemen," a quartet of behaviors that almost invariably lead to the downfall of a couple: criticism (as opposed to simply voicing a complaint), contempt, defensiveness, and stonewalling.[76] The Gottmans have also discovered that creating and sustaining a happy relationship is much more complex than that, and I would like to argue that this is because most people . . . not just those in romantic relationships, but in *any relationship* . . . approach conflict as an essentially *unconstructive* activity. We tend to see it as an adversarial showdown in which our only goal is to smash the other person's point of view to preserve our own. Nothing constructive at all . . . no sense of exploration or entertaining new notions, no openness to "travel" or new experiences, no eagerness to discover if there are perhaps implicit and harmful assumptions that need to be made explicit. No questions, just "answers" that lead to an uncreative, uncaring, unintentional kind of deconstruction. But what if, instead, we went into every conflict with a completely different mindset . . . perceived conflict with very different assumptions . . . namely as a chance to discover another's deviance? Since to understand another—even oneself—is not to know how one is similar to others. To love is to love another's deviance. What makes you *you* and them *them* is how one deviates from the norm. As my partner, Isabel, and I like to say, it's about whether two people's madnesses are compatible. To enter conflict consciously with doubt as a part of your ever-evolving history of perception is difficult, risky even, especially if the other person is not entering the conflict with an understanding of the principles of perception. Thus, in addition to humility, creativity requires courage because you are stepping into a space that your brain evolved to avoid.

How do I step into a place that evolution told me is a pretty bad place to go—the dark of uncertainty—by celebrating doubt? How do I turn all this new understanding into a way not just of seeing but of being? What is the first active step to deviation, to questioning my assumptions in order to go from A to B? To borrow a phrase from an old Bob Newhart sketch that I love . . .

JUST STOP

I mean this literally. Not slow down, but stop.

As noted in Chapter 8, if you want to go from A to B, then you must actively engage with the world. But the first step to get to B is to go from A to *not-A*. To be in not-A is to be in uncertainty, to experience the stimulus without the requisite meaning of the past. The key is to choose to look away from the meanings we have been layering onto stimuli. Stop your reflexive response with awareness . . . as one can do when one is able to see the cause of a reflex. Someone bumps into you on the street. Your first automatic response could be: *What an asshole!* That is "A." But just stop. Don't go to A. Go to *not-A*. Perhaps the person who bumped into me is ill; that's why they stumbled, and they may need help. Or they may indeed be an asshole. Don't know more. Stopping gives you the chance of knowing less, of halting the perception-narrowing force of the cognitive biases that we are always trying to confirm, of taking the jerk out of the knee-jerk and sitting with the meaninglessness of the stimuli, even if it doesn't feel meaningless.

When all is "happy," there is really only one choice: stay with meaning A. That's obvious. There really is no choice at all . . . unless one actually finds reward in destruction for the sake of it, which, sadly, some do. But for most of us it's only when the situation is in conflict that we truly have the chance to exhibit choice. One might even argue that it's in conflict that our character is revealed *and* created. One option when in conflict is to jerk the leg out, as one always has in the past. That is a normal (i.e., average) response. But now that you've awareness of why you see what you do, you have another option: to deviate . . . to choose not-A . . . to enter uncertainty. That is where control lives, not in what comes after not-A, whether it be B, C, or Z. But to *look away* from the obvious, from the well-entrenched attractor state is hard. First, you must stop your first response.

As you begin to do this, you dim the intensity of the influence your current assumptions have on your perceptions. This is essentially what meditation is . . . a mechanism for going to *not*-A, the goal of which is for you to "empty your mind" and stem the ceaseless (and often fruitless) flow of meanings. Apart from helping with amygdala fight-or-flight responses to stress, meditation has been shown to help people develop more empathic thinking (empathy is nothing if not a creative process of imagining yourself as another person) and thought patterns that involve "mind wandering." A 2014 study at Harvard even showed that an eight-week mindfulness course increased the gray matter in the participants' brains, proving that meditation itself is an inner enriched environment that fosters neural growth.[77] In stressful situations in which creativity often loses out to your fight-or-flight response drives, just stopping actually engages a different chemical in your brain than the distress-induced cortisol. It releases *oxytocin*, which is quantifiably associated with more empathic and generous behavior (as well as other behaviors). Through generosity and empathy we listen much more creatively. So, you can actually create better living through your own chemistry.

Stopping our reflexive meanings is so important because it is the first step into uncertainty from which we can create new, unforeseen meanings, and in doing so re-create the meaning of past experiences through the tool of delusion, which alters your future past, thus in turn altering your future perceptions. Free will lives, not in going to A, but in choosing to go to *not*-A. Free will lives in changing the past meanings of information in order to change future responses. Free will requires awareness, humility, and courage within the context of uncertainty.

Several years ago there was a period in my life that was particularly and continually stressful. The result was a moment in my life when I "crashed" emotionally and physically. My body entered a state of fairly constant illness: Headaches, spontaneous numbness, a large number of neurological symptoms. This is a dangerous thing to experience as a neuroscientist, both fascinating and fear-inducing at the same time, as you know too much and nothing at all. Each symptom (stimulus) triggered a series of assumptions and consequent perceptions, from brain tumors to MS. I went through the lot for weeks, if not months. Eventually, as there was no clear diagnosis despite the symptoms being not only psychological but also physiologically measurable, panic attacks ensued. The associated perceptions were as real as with any other perception. The sense that I was dying—immediately—was profound. It was so profound that ambulances would be called, which upon their arrival became a necessary step from the unknown to the known, and thus the sense of impending death would pass. So how did I recover and get from A (feeling awful, with phasic panic attacks) to B (feeling a new "normal," whatever that is)? Like many, many before me: by embodying the process of perception and applying the essential first step to moving in a new direction, becoming aware of my biases . . . and with that understanding, stopping the movement in the current direction. I just stopped my speculative worries. I *actively* ignored them. And in that suspension of thinking, new and much better thoughts began to appear. This ignited a new history for my perception.

When it comes to anxiety attacks in particular, one of the routes . . . if not *the* best route . . . to shifting away from them is to ignore them. And I mean to truly ignore them. As the famous psychotherapist Carl Jung said, problems are never fixed; we only change our perspective on them. In this instance, the change in perspective was to not pay them any attention at all. Not to try and figure out their cause, as doing so actually reinforces the meaning, making subsequent attacks stronger. It was easier to be in A ("knowing" that something was wrong) than in *not*-A (I was experiencing a useless delusion). It was difficult to look away, and as studying attention demonstrates, it is a challenge all humans experience from a very young age.

When my second son, Theo, was a baby, one day he was sitting in a bouncy chair on the kitchen table in our Stoke Newington house in London. He was perfectly happy because suspended in front of him were four highly contrasting soft toys that were just out of reach. Every time he would bounce in his seat, the toys would bounce in response, resulting in what initially seemed like endless enjoyment. There were only smiles and even laughter while I cooked nearby. But then something changed. He started to cry (not that surprising, as that's what babies do a great deal of the time). So I did the obvious thing and made the toys move again. It drew his attention immediately, and yet he continued crying.

It was sad, but, well . . . it was also fascinating. I observed rather quickly (yes, I watched him as he was crying, but his cry wasn't desperate yet! . . . one of the various costs of having a perceptual neuroscientist as a father) that his crying would slow or stop as his eyes moved away from the toys. But then, as if drawn like a magnet, his eyes would pop right back to attending to the central toy, at which point he would begin crying again. He would then try to move his attention away again, ceasing his cry in the process, only to have his eyes bounce back again to the central toy, and he'd cry yet again. It was as if his crying was out of frustration, I thought, taking all of this in. But frustration about what?

That the toy had more power over him than he had over himself.

I realized that Theo no longer wanted to look at the toy, but he couldn't help but do so. It was remarkable. He couldn't stop responding. He couldn't move from position A. He wanted to move to a position of not-A, but was frustrated with his own inability to divert his attention. So after three hours of watching this behavior, I finally removed the toy and all was well—*todo bueno!* (Just kidding: I removed the toys very quickly.)

What does this little personal experimentation on my own child tell us? While we still know very little about attention (which can be a convenient catchall for neuroscientists), it seems that the power of attention is not in doing the looking, but in the ability to stop looking . . . to look away, to move your eyes toward the less obvious, to stop a cycle of thoughts and perceptions. This is because our attention is naturally drawn to things that historically mattered to us, whether they be as basic as high-contrast objects in the case of young, developing visual systems (which is like an insect's visual brain), or stimuli that were associated with pain or pleasure.

Clap your hands near an unsuspecting person, and they will immediately divert their attention toward your eyes. At a party, stand next to your friend who is talking to someone else, and purposefully start telling the person you're talking to about your friend, saying their name audibly. Notice how quickly your friend will either join your conversation, even if you stop talking about them, or move away, simply because they find it nearly impossible to continue their current conversation. Or consider the thousands of hours we spend in traffic on motorways that are not actually blocked by any accident in our lane, but because the accident in the *opposite* direction draws the (morbid) attention of the drivers in the opposing lane. This power to stop is the power of the frontal cortex of your brain. The point of attention is not in looking toward, or in directing your gaze to the rational . . . the things that make most sense statistically and historically, as in the case of Theo. Rather, the power of attention is in looking away from the "obvious," toward the less obvious. The power of attention is in beginning deviation, and steadfastly challenging your brain to engage the frontal-cortex-driven inhibition of stopping when needed.

In this and pretty much every other regard, we're very similar to other animals, perhaps more than many want to believe. Not only do bees with their mere one million brain cells see many of the same illusions we see (and see in ways that our most sophisticated computers are unable to), but birds can think in abstract ideas, and certain primates get upset if situations aren't fair. So, too, with the directionality of attention: Much like other animals that are drawn to shiny objects, we too attend instinctively to what is shiny (i.e., high contrast both physically *and* conceptually). Our eyes move systematically, rarely randomly. We look at borders between surfaces. We change our eye movements according to our emotional state, looking more at the foreground when we are in a state of higher power, more at the background when we are in a state of lower power. If you're a woman, you're more likely to look at another woman's face differently than if you're looking at a man's (at the mouth instead of the eyes, for instance). Our eye movements, which are our windows into the world, belie our assumptions. Thus, to look at the obvious is an obvious thing to do, though each of us has a different "obvious." And the narrow focus of what you look at can determine the meaning of the rest of the information.

Thus, to look at the obvious is an obvious thing to do, though each of us has a different "obvious." And the narrow focus of what you look at can determine the meaning of the rest of the information.

You'll have by now noticed that in addition to the diamond at the bottom-right corner of the book, there is another pattern of lines in the bottom-left corner. Here, I want you to flip the pages.

Like the diamond, the bars can appear to move in two different ways. You can either see the two bars moving left to right and back again in the form of an X (which is called pattern motion), or you can see them as two independent bars moving up and down *separately* from each other (which is called component motion). It might seem that they flip from one direction of motion to the other randomly. But they don't. It has everything to do with where your eyes are looking. To illustrate this, I want to focus your gaze on the center of the X, where the two bars meet. When you do this, you will see an X moving left to right. When, however, you fix your gaze on the end of one bar, and follow its motion up and down, you will see the two bars moving independently up and down against each other. What you see is determined by where you look. This is because, while the entire image might have two equally likely possibilities, different elements of the same overall stimulus do not. The space of possibility is determined by where you look, not by the abstract possibility of the image itself. So what does this reveal?

Fundamental to defining a person or collection of people is not only what they do (i.e., what they look at), but also what they don't do (what they don't look at). In the brain, for instance, it's not only the active cells that determine the nature of what we do, but also the inactive cells, since it's the overall pattern of activity that matters. Thus, unlike most animals, we can change our future past by looking away from the obvious . . . by going from A to not-A. What we look at reveals us, but what we choose to not look at creates us.

What we look at reveals us, but what we choose to not look at creates us.

It's in looking away that you begin the process of deviating, because not looking at one thing means your eyes must now settle on something else. It's that "something else" that you've much less control over, as your brain will now settle into a new attractor state given the new stimulus combined with your past history. But at least it won't be A. Do this enough, and in the future you're less likely to reflexively see A, and more likely to see B (or C, or Z). We can generalize this process: What is true for how we direct our eyes is also true for how we direct our thoughts. It just takes practice to guide them.

Let's go back to the flipbook of moving bars now, and do it again. But this time, you'll have noticed that there is a dot just above the intersection of the two bars on each page. I want to you keep your gaze fixed on that dot as you flip the pages. While you do so, I want to you focus the *direction of your attention* on either the junction between the bars or where one of the bars meets the boundary, just as you did previously with the movement of your eyes. This time, instead of moving your eyes, *move your attention* only. Notice how again the bars will change their direction of motion depending on the direction of what you attend to—and don't attend to. What we look away from (and toward) changes the nature of the information we encode in our brains and the meanings that we see. Thus, we can shift from A to not-A by "looking away" both externally with our eyes and internally with our "minds." We can rewrite our perceptions through stopping, then beginning anew. Being in not-A . . . and all that is required to get there is what I call "ecovi." But this requires going to a place that our brain evolved to avoid . . . namely, uncertainty.

CHAPTER 10

The Ecology of Innovation

We have seen that understanding how perception actually works opens up a whole new frame for looking at creativity and creation more generally. By learning *why* our brain evolved to perceive the way it does, we are able to engage ourselves with steps that can change the way we see. This approach illustrates the role of the individual, since we are each responsible for actively questioning (and the consequences of doing so). But, as I have emphasized throughout this book, none of us exist in a vacuum. We not only create our ecology . . . the environment we inhabit that is constituted not just by the things in it but also by the *interactions* between them . . . we are created by it as well. Thus, to truly learn how to deviate, we must learn how to innovate our perception within the world with which we interact. It is here that we can discover a new way of being for ourselves and others, and construct a place that ties together all of the perceptual principles we have examined.

At the end of a dimly lit hallway on the fifth floor of the Valley Life Sciences building at the University of California, Berkeley, stands a nondescript door. It promises very little intrigue or surprises as to what's on the other side: a janitorial closet, perhaps, or a storage room full of office supplies. In reality, behind the door is a both strange and revolutionary space within four large adjoining rooms. A latticework of struts and supports hangs from the ceiling, connected to electric cords, wiring, and flood lamps. Computers and video cameras abound, along with engineering tools and circuit boards. Most striking of all, on nearly every surface sprawl small robots that look like insects or reptiles . . . not just a bit like them, but remarkably so, down to the joints of the limbs and the sinews of the bodies. When I step inside a narrow room and ask a Ph.D. student what brought her here, she answers simply, "Cockroaches."

This is the Poly-PEDAL Lab of the biologist Robert Full. PEDAL stands for Performance, Energetics, Dynamics, Animals, and Locomotion, yet even these five words strung together in such a geeky-cool way don't tell the "full" story of Full's lab or its wonderfully eccentric and stunning achievements. Full and his sundry collaborators, from undergrad college students to renowned experts in wildly diverse fields, have excited scientists and nonscientists alike with their discoveries. The Poly-PEDAL Lab has solved the millennia-old mystery of why geckos are able to stick to walls: by using van der Waals forces, an incredibly complex rule of intermolecular attraction and repulsion. They have answered the question of why cockroaches can flip around at top speed and keep moving upside down: hind legs with adhesive hairs that hook onto a surface and let them swing into a "rapid inversion." And they have uncovered how weaver spiders move efficiently across mesh surfaces that have only 90 percent of the surface area as compared with solid surfaces: through a foldable spine that allows them to spread contact onto different parts of their legs. The Lab has asked some of the greatest and (in hindsight) most insightful questions about the most inexplicable phenomena of natural biomechanics ... and answered them, too. Their answers aren't just interesting in and of themselves; they also reveal a powerful principle at work.

Full's goal isn't *just* to unlock these mysteries. As the Lab's credo states about the critters whose bodies they model, "We do not study them because we like them. Many of them are actually disgusting, but they tell us secrets of nature that we cannot find out from studying one species, like humans." Full's goal is to apply these "architectural secrets" to human endeavors through advances in robotics, thus reverse-engineering (up to a point) facets of the creatures whose movements he and his labmates study. He has had nearly unprecedented success doing this, essentially creating the new field and design philosophy of *biological inspiration*, as well as other subfields such as *terradynamics* (how things move on surfaces, as opposed to aerodynamics) and *robustness* (how structures achieve their most robust form). One innovation that came out of his lab is RHex, a cockroach-inspired robot that can maneuver terrain better than any previous experiment in bionics. While all its applications are still being explored, it is already beginning to change the literal frontlines of conflict zones. The American military has deployed it in Afghanistan, reducing violent encounters by sending in RHex as the robotic vanguard to clear areas before soldiers go in themselves.

In spite of the parade of breakthroughs his Poly-PEDAL Lab has produced, Full himself is a humble, unprepossessing man who values collaboration over ego, and exploration over prestige. With a full head of silky white hair and a charismatic walrus-esque mustache, he isn't driven by usual ambitions. He is like a brilliant, caring uncle, as gentle as he is wise and experienced. As he was one of my supervisors when I was an undergraduate at Berkeley many years ago, I know this firsthand. "It's in the curiosity," says Full.

As the head of the lab, I suggest Full's job is not . . . as was once the case for leadership . . . to have all the answers. Rather it's to have the good questions. Thus, Full's first job as a leader is to create a sense of caring (often through inspiring wonder—though there are other routes), since if you don't care, you'll not have the necessary drive to face uncertainty and the momentum of neurological and cultural attractor states discussed previously. Second, is to ask Why? of just the right assumptions (which are often hidden from sight) . . . and then How? or What if? . . . and then to observe . . . and finally to share . . . and then to repeat this process again and again. He's clearly extremely good at this. Asking great questions that disrupt is not enough, however. A great leader must also create a space for others to step into uncertainty and to flourish, since he understands that the success of the system is determined by those he leads and how he leads them into uncertainty. This is why he chooses to work with people who will welcome the ways he disrupts conventional thinking in order to create new opportunities. (And he doesn't get defensive when faced with people who disrupt his own thinking . . . on the contrary.) Periodically, he enables the collective to just stop. As such, he has created a lab based on the **solution evolution itself has given us to uncertainty**.

What is this solution?

Answer: It is *a way of being* that can profoundly change how we live . . . a way of being that applies to most things innovative and pathbreaking, a way of being that is defined by five principles that form the backbone of this whole book:

I. Celebrating uncertainty: to approach "stopping" and all the questions this stopping spawns from the perspective of gain, not loss.

II. Openness to possibility: to encourage the diversity of experience that is the engine of change, from social changes to evolution itself.

III. Cooperation: to find value and compassion in the diversity of a group or system, which expands its space of possibility—ideally, combining naïveté with expertise.

IV. Intrinsic motivation: to let the process of creativity be its own reward, which will enable persistence in the face of tremendous adversity.

and

V. Intentional action: ultimately, to act with awareness . . . to engage consciously, from the perspective of why.

Remarkably, principles I through IV are defined by one word: **PLAY**. By "play," I don't mean so much a literal activity as an attitude. It's about embodying playful qualities in how one approaches a problem or situation or conflict.

The Oxford natural philosopher and primatologist Isabel Behncke is a specialist in the science of play, with an emphasis on adult play, specifically among bonobo apes. She has lived with bonobos in the Congolese jungle for extended periods, closely observing their habits and behaviors. Bonobos, along with chimpanzees, are our closest living relatives. But unlike chimps, they are unique in that they are highly sexually promiscuous (including male–male, female–female, adult–infant, etc.). These adapted traits are useful; sex is used as a language to regulate conflict. But while bonobos are well known for their sexual promiscuousness, Behncke discovered that play is in fact more frequent in their primate society.

Humans are very much like bonobos. It has allowed us to engage with the world and learn *because in play uncertainty is celebrated*. If you take away uncertainty, the game ceases to be "fun." But fun does not imply "easy." To play well is hard (as any Olympian can attest).

What is more, unlike most activities to which we humans and other species such as bonobos devote our energies and hard-earned calories, Behncke and others have shown that play does not have a post hoc nature (*post hoc* means that what matters is the results the activity produces, not the process itself). Post hoc activities include hunting (result: food), working (result: ability to pay for food and shelter), and dating (result: sex and/or romance). But play stands out as a unique pursuit in that it *is intrinsically motivated. We play in order to play,* just as we do science to do science. It's that beautifully simple. **The process is its own reward.**

A further, crucial aspect of play is that we often do it with others, not just by ourselves. When you play with someone, it has a lot to do with who you are and who I am, be it playing racquetball, poker, or sex. But I might play differently with someone else than I do with you. If, however, I played with you as if you were someone else, or with just a generalized idea of what play is, this would distort and limit the things we could authentically do together. Yet many misguided brands speak to all customers as if they were one average customer, so that in the end they basically speak to no one. Treating people like the average playmate, then, ignores deviance. Research on play has shown that it is a way of safely learning other people's deviance. Play is a way of engineering experiences that reveal assumptions that we then question, unleashing unpredictable results. The underlying premise of Behncke's work is that play makes your system more complex, increasing the likelihood of inspiration.

While play enables one to step into uncertainty and thrive, play alone isn't a complete tool, as much as a childlike approach to life can be generative for the brain. To survive during evolution, innovation needs more than creation. We also need principle V: Deviate with *intention*—not just for the sake of deviance (though there can be value in that, too, in terms of a random search strategy). This is essential, and we have a clear example of its significance. What do you get if you add intention to play?

SCIENCE.

This word will evoke in you a series of assumptions . . . most of which, I fear, won't be terribly positive. You're likely to be thinking that science is about the sterility of lab coats and unimaginative facts. You might be thinking that science is a methodology defined by and through measurement, and is thus the ultimate in information harvesting. This is not what defines science. Nor is science defined by its "scientific method," which is only one expression of a way of perceiving and acting that is in fact much deeper. To be sure, the skills needed to design and run *a good experiment* are essential and can be very difficult indeed to learn and hard to run well: There are few things in my view as elegant or beautiful as a well-designed experiment, in the same way that there are few things as compelling as a beautifully executed painting, song, or dance. But just as with art, so too with the craft of science. The craft of the medium doesn't necessarily define it.

Evidence that this way of being has practical application is found in a class of ten-year-old children (including my son Misha) who in 2011 became the youngest published scientists in history after conducting a study of the perception of bees. By applying these principles of science ("play-with-intention"), the Blackawton Project, as it was called, challenged the world of science as the editors of many, many journals rejected the children's work (even the largest funding institution for public engagement in the UK . . . the Wellcome Trust . . . declined to fund the project, arguing that it wasn't likely to have a large enough impact). Yet the program (we call the iScientist programme, which was devised with the brilliant British educationalist David Strudwick), created a wholly new space of possibility that had not been explored before . . . the process of innovation that *unites* creativity and efficiency rather than pursues them independently.

Together, creativity and efficiency define innovation. Innovation *is* this dialectic that is present throughout nature and is echoed in a number of other dualistic tensions that run through this book: reality versus perception, past usefulness versus future usefulness, certainty versus uncertainty, free will versus determinism, and seeing one thing but knowing another is actually the physical truth. This elemental dialogue between creativity and efficiency is embodied in the brain itself, which is perhaps the most innovative structure in existence.

The human brain is balanced between excitation and inhibition. This is essential, since it keeps your brain at the point of maximizing its ability to respond: too much inhibition, and stimuli will not propagate properly, yet with too little inhibition a stimulus can create an excessive closed feedback loop that results in an epileptic fit.[78] Increased use changes the balance of excitation, and as a result the inhibitory connections must grow in order to maintain the balance (and vice versa). This enables the brain to maintain a state of readiness in any environment . . . to be able to respond to change in uncertain contexts . . . and this is why the brain is effectively matching its complexity to its context. The brain is adapted to continually redefine normality . . . to constantly look for a *dynamic* equilibrium. It creates a discursive, ongoing process of growing and pruning, growing and pruning. This is simply the brain moving back and forth between creativity and efficiency, and between the increasing and decreasing of the dimensions of its search space. Recently, neuroscience has discovered two larger cellular networks. One is the "default network" that is more active when the other is resting and "free-thinking." The connectivity of the default network is more vast and inclusive. The other is more directed and efficient, and tends to be activated during focused behaviors.

To simplify things into a straightforward rule of thumb, think of it this way: In an ecology of innovation you can frame creativity as saying yes to more new ideas. Recall Chevreul and how he solved the mystery of the Gobelins fabrics. If the king had given him just one month, or even one year, to figure out the problem, he would have failed. The imposition of efficiency would have curtailed his lengthier yet necessary investigations. But since Chevreul had a timeline that allowed him to eliminate erroneous explanations (for example, regarding the quality of the tapestries) and explore more counterintuitive explanations, this eventually produced a tremendous and historically important jump forward in our understanding of human perception.

While we love to worship "geniuses" like Chevreul, Steve Jobs, and their ilk, sometimes the efficiency of saying no is also essential. Thus, efficiency in and of itself is not a bad thing. Saying no, yet not establishing a Physics of No, is its own creative art. The people who know how to properly implement and work with exploratory creative minds are essential, like editors who set deadlines for writers and help them tell their stories in the most compelling fashion possible (as Arron Shulman has done for me in this book). After all, sometimes efficiency will save you more than creativity. Think of the bus coming at you and how useful your fight-or-flight response is at that moment. You don't want to ask yourself, "Hmmm, is there a way I could see this differently?" Because the answer is, yes, there is. But you probably shouldn't try. The wise decision is to get out of the way as efficiently as possible. This gives you a better chance of surviving—which is always the point.

All things being equal, then, a system that consumes less energy (literal or financial) in completing a task will outcompete another system that consumes more, whether in the context of biological evolution, business, or personal development. As such, a focus on efficiency has been the approach of most industries.

Even schools and universities . . . the very places that are meant to be there to inspire different ways of seeing at the individual, cultural, and social levels . . . have come to be incubators of efficiency. This is, of course, deeply ironic: "Businesses" are there explicitly (even legally) for themselves, not for the people who are their human manifestation, and the companies are (mostly) straightforward about this. So we—to a certain extent—accept it (except in situations where we experience hypocrisy; the brain is very sensitive to authenticity). But surely the point of schools and universities is to be hothouses of great questions. Great questions can't be "produced" on an assembly line. I can't say to my Ph.D. student or postdoctoral fellow, "I'd like that discovery by Tuesday, please!" Nor can the value of a discovery be quantified in terms of dollars (at least not in the immediate future), making it wholly wrongheaded to attempt to maximize creativity by ratcheting up efficiency. Yet this is precisely what happens with creativity in education: it gets crunched into a competitive economic model. This is like seasickness: a contradiction of the senses, resulting in not just intellectual but also human costs. Incredibly sadly, in 2014 a professor at Imperial College London committed suicide. His university was threatening to remove him if he didn't bring in large sums of grant money.[79] While we can't speculate about what other struggles he may have privately faced, universities are increasingly pressuring faculty to achieve goals that have little to do with the true aims of education: teaching and expanding our intellectual and scientific universe. Surely there needs be one place in the world where one purposefully runs a deficit, simply because creativity is inefficient in the short term (but not in the long). But instead that kind of deficit spending seems to be reserved for the least creative of our institutions (governments). And yet, running on the efficiency model of businesses (even without the financial infrastructure of a top-flight business) will reduce the creativity arising from universities. As a result, the work will shift toward translational research, with ever fewer foundational discoveries that will reap rewards in the long term being made. It's interesting, then, that consistent with this perspective, a paradigm shift is occurring between universities and businesses, wherein more creative work is starting to come from the latter, in the names of Google, Facebook, Apple, and the like.

(*end of rant*)

Given such a ubiquitous focus on maximizing efficiency, it is worth asking whether there is a best environment for this practice outside of life-or-death scenarios. Biology—including neurobiology—gives us an answer: *a competitive environment.* Competition is a great motivator and process for eliminating "the fat" . . . the superfluous, the redundant, the outdated. But evolutionarily speaking, there is a lot of death in competition. As a business model, competition means that a company will have a culture of termination, in which people are fired for not producing certain results according to a given timeline. While such an approach can spur extreme effort under pressure, it is exactly the wrong way to promote creativity . . . not just because exploration is not prioritized, but also because employees' stress levels will narrow their spaces of possibility.

Even at the level of the brain we see evidence of maximizing efficiency through competition. As you may recall, 20 percent of the energy that your body consumes is spent (or should be spent) on this 2 percent of your body mass. So, brain cells cost a lot in terms of the biological material required to make the trillions of connections, as well as generating electrical activity along those connections, hence the energetics of simply *thinking* (which doesn't necessarily mean creative thinking). This is why your brain tries to minimize its number of brain cells and reduce the number of times each fires. The process plays out like this: Either you have lots of brain cells and each one fires only once in your lifetime, or you have only one brain cell that fires continuously throughout life. It turns out that the brain employs *both strategies*. It balances the number of cells with requisite activity by producing growth factors that compete for limited resources of energy. Those that are the most active . . . or, more accurately, are the most active concurrently with the cells that they connect to . . . are more likely to survive. However, the amount of available resources isn't so limited as to eliminate redundancy in our neural circuitry, which is essential for creativity and resilience.

Think of the cerebellum of a bird as compared to the wiring of an airplane. The bird's cerebellum controls its motor behavior, as does the cerebellum in mammalian brains. In the case of the bird, it enables coordinated wing movements that are essential for fighting the laws of gravity. In contrast to ground-dwelling creatures like ourselves . . . who, if their motor coordination fails, might fall a few feet at most, with the requisite cuts and bruises . . . a failure of the bird's cerebellum would be catastrophic. As such, you might imagine it was very efficient indeed. This is true. But it's also true that a bird's cerebellum has a tremendous level of redundancy of connections: it is theoretically possible to remove over half of it without causing the bird to tumble to earth![80] Compare this to the most sophisticated warplane, which is possibly one of our most efficient creations. Cut just a few wires in its control system, or damage one of its wings just a small amount, or even have a bird fly into one of its jet engines, and hundreds of millions of dollars will crash to earth.

The system, like most businesses (and indeed much of modern-day life, from sports and athletes to education), is highly efficient but not at all creative, and as such is not adaptable—**i.e., innovative!**

To create a successful ecology of innovation we must look to our own biology and tap into our very own neural nature that balances efficiency and creativity. This is why neither always being creative *nor* always being efficient is the end-all route to success. The two qualities must exist in dynamic equilibrium. What is more, the system must develop.

The process of development is the process of adding dimensionality, called *complexification*, which my lab and others have studied for many years. It's very intuitive: start simple (few dimensions), add complexity (more dimensions), and then refine (lose dimensions) through trial and error . . . and repeat. Development is the process of innovation incarnate.

In 2003, when Apple began work on its revolutionary smartphone, they were sure that they wanted to use new "multi-touch" technology for the screen, which would radically change the user interface experience. But design of the phone itself was still unclear. One approach was called "Extrudo." The case was made out of extruded aluminum, with a fatter middle and tapered sides. Not totally satisfied with the Extrudo, however, Jony Ive also spearheaded another approach, the "Sandwich," which combined two halves of specially designed glass. Ive and his team were complexifying, simultaneously entertaining the possibility that both of these assumptions were good to help them more creatively discover which, in fact, wasn't. In the end, neither turned out to be the chosen solution. Instead, Ive decided on a previous prototype that they'd had all along . . . a third assumption that his team wouldn't have chosen if they hadn't first complexified by designing the other two.[81]

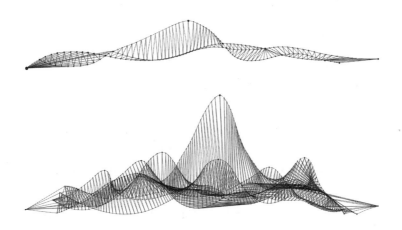

This complexification model is effectively the Silicon Valley model, though it has its roots in biology, since development, learning, and evolution are *the* iterative processes. So, "innovative" Silicon Valley are companies doing nothing new. They are basically doing what slime molds and brains do all the time. There are a few important tricks and strategies to ensure that complexification succeeds, though, which David Malkin and Udi Schlessinger's inspired Ph.D. work in my lab revealed.

Thinking back to modeling assumptions as a network, we have shown that the more complex the network, the more likely the "best" solution will exist in your search space of possibility, simply because the large number of interconnections increases the number of potential solutions (as shown in the bottom image here). This is in contrast to very simple systems that have only a few possible solutions (as shown in the top image). The problem is that complex networks don't adapt well (this is what we call *low evolvability*). So while the highest peak is more likely to exist in your space of possibility if you're a complex system, an evolutionary process is unlikely to find it, which creates quite a conundrum: complex systems can help us adapt, but aren't very adaptable themselves.

So how do you find the best solution?

You develop.

Start with one or a few dimensions, then add more. The image here shows the progressions of the same system as it develops. David Malkin and I call it an "uber-landscape," which is a unique idea that adds the element of time to a search space (which are usually modeled as static). Notice how the system starts off very simply . . . with a single peak. But as time progresses along the z axis, the number of peaks steadily increases as more elements are added to the system, until it reaches its most complex state at the other end of the plot of the graph, where there are roughly five

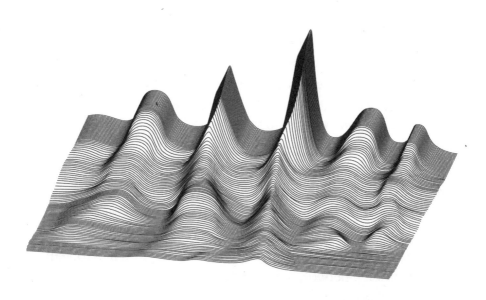

peaks. What we found surprising is that when one grew the network in this way, the best solution in its most complex state was also the solution that had the highest probability of being discovered . . . when at each step the complexifying system tried to minimize its "energy state." This is in contrast to our search for the same solution when the system started in its complex state in the first instance. It's as if, when starting from the simple state, we were able to walk along a ridge from lower to higher dimensional space. In doing so, we were able to walk past the troughs of a lower-dimension system by adding a higher dimension, which is called *extra-dimensional bypass*. The findings suggest a counterintuitive, but essential, and admittedly speculative principle: Adding what we call "noise" (i.e., a random element) to a system can increase the adaptability of that system. If true, then—though highly speculative—there could be genes that are there not because they encode any particular phenotype, but because they increase the dimensionality of the search space *full stop* . . . and are then lost (or silenced) once the system settles on a new peak. Imagine what this would mean in an organization where the only function of a particular person is to increase the dimensionality of the space, thereby bridging the poorer solutions . . . suggesting that sometimes . . .

Noise in a system is good!

Having said that, it is very important when adding new dimensions that you balance the rate of periodic success with the rate of complexification . . . while minimizing the energy state of the system at each level of complexity. In other words, you wouldn't want to add updates and upgrades to a fundamentally crappy cell phone; you should only do this for one that is already strong at its core. This mirrors how the brain develops. When it grows, it is adding connections, and therefore **complexifying itself**: adding dimensions to the space of possibility, and forging new pathways of neuroelectric connectivity. But it's doing so according to feedback on what worked and didn't work in the past. The internal patterns of activity that worked (the useful assumptions) are reinforced, and those that didn't work (the no-longer-useful assumptions) are lost. As you may recall from Chapter 3, this is what your cells do in the neuromuscular junction, creating redundancies that the system then prunes away, only to then complexify again, with the result that the number of connections is actually greater overall, but organized usefully. What is more, Udi Schlessinger also demonstrated in the lab that the process of useful complexification has its own useful progression. Rather than just add a random connection, it was more useful to duplicate the system (add redundancy) with an additional small step. This process is something that deviators embody not just in their cells but in how they live their lives . . . complexifying their existence with new experiences that complexify their assumptions, one small step at a time . . . and in turn alter perceptions, not just in their minds but also in the physical world, only to then follow this process of complexification with a process of consolidation. Each update, then, is a "riff" on the last version, not a rethinking of the whole thing. This cycle of increasing and decreasing complexity IS the innovation process that is endemic to life itself, and the underlying route to resolving its inherent conflicts.

So, if you own an iPhone or any other Apple product, there are always new modifications, new updates, and new models. This is not just about marketing as such, but a result of the questioning process that Ives pioneered through the design group at Apple—the governing idea that, even though the company has an unequivocally great and successful product, they must keep asking themselves, *Why can't we make this better?* They can and do, continuing to unloose new little "jokes," as embodied by each new iteration of the iPhone that improves on the last.

In addition to brain development, evolution is also a great example of the successful tension between these two poles. All species go through exploration (creative) periods, in which different traits appear, like feet in the first organisms that left water billions of years ago, or like the opposable thumbs that developed in some primates, including humans. There have been periods of massive expansion of the diversity of species (increasing dimensionality), followed by the exploitation (efficiency) period in which those with useful new traits survive (decreasing dimensionality) while others are selected out, leaving only the most robust organisms.

We can also see this cycle reflected in the movement of Earth in relation to the Sun, and how humans evolved to be in synch with their environment. If our species' existence is itself an ecology of innovation, then the balance between wakefulness and sleep is at the heart of the process, where wakefulness creates connections and sleep consolidates them. The same goes for our development, first as babies and then as children. When we are young, this critical period for our brain is plastic . . . as it's *establishing* many of its future assumptions and attractor states. This is the tremendous advantage of our species being born "too early," as small beings that can't survive on our own for many years, with brains that are underdeveloped. Because of this, the human brain adapts itself to the environment in which it happens to find itself. It is for this reason that humankind, more than any other species, has been able to occupy such a diversity of niches in nature. Then, as we age, this formative period slows and change becomes more difficult, as our repeated experience forms deeper attractor-like states (though of course we can still creatively change . . . that's why you're reading this book!). Yet in allowing you to be creative, these "detours" enable you walk along ridges around the valleys between different peaks in a fitness landscape. Once you end up on the higher peak, you can discard that extra dimension and swing back toward the efficiency side of the paradigm. Innovation (adaptation), then, is like a spiral: when you're in the cycle usefully, you never return to where you were. Instead, you return to a similar but higher place.

A useful metaphor is the tonic scale in music. When you're at middle C, as you move up through notes D, E, and F, you are moving further "away" from C in that the signal is shifted to higher frequencies. This is akin to the creative phase. But  as you continue still further up the tonic scale, through G, A, and B, you start moving *perceptually* "back toward" C, but one octave higher. This spiral nature of innovation illustrates that creativity—ironically —is a key route to efficiency, and efficiency can lead to creativity. Like all things in nature, it's the movement between them that matters in order to create an ongoing process adaptable to a changing world.

This is why in business and other cultures that attempt to foster adaptation (which should include universities), the key is to never be "phase-locked," wherein one pursues a single idea with maximum efficiency without movement. In nature, as in business, such systems are quickly selected out of the population. The app called Yo was a single, highly specialized and very efficient messaging system that reduced communication to a single word: "YO!" The significance of the message varied with context. It took the founder only eight hours to code, yet it quickly won over hundreds of thousands of users. Nothing could compete with it. It was Malevich's *White on White* of messaging platforms, taking the genre to its ultimate conclusion. But had it not complexified, it is likely to have quickly died, or even more likely to have been subsumed by another, more general platform. Indeed, the most successful companies follow what I call *wedge innovation*. They start off "sharp," with a highly focused product that enables them to outcompete others in their space (by being efficient with an original, creative idea with which they started). But then they widen their use-case and thus the company's contextuality, increasing what they do and why they do it. Kodak is the classic example of a company that did not do this and died because they didn't expand their wedge (the pioneering of print photography for everyday people) to start a further twist to its spiral by expanding into digital photography before their competitors, although they had the opportunity to do so. Apple is an example of the opposite . . . a spiraling, forward-thinking company.

As a company grows (and indeed as life grows), it needs to have different spirals simultaneously underway with a variety of products at a variety of time-scales (frequencies). For example, when Google reaches the efficiency phase of one of their products . . . that is, it works well, like the Google search engine or the Chromebook laptop . . . the company is already in the beginning phases of other innovations, like self-driving cars. This could be considered the point of GoogleX, Googe's moonshot research and development initiative, which is to create the creative ideas that others then make efficient (which I'd suggest is the same process that a lab follows . . . as exemplified by Bob Full above). The result is that some ideas will become innovative and others will die, like Google Glass .. not because of the technology, but because Google didn't seem to consider one fundamental attribute in the design: human perception/nature. Google doesn't appear to have taken into account the importance of eye-gaze in communication. We move our eyes not only to obtain information but also to signal our state of mood and the nature of the relationship in which we are engaged. For instance, if I'm looking at you and you look down, you may be signaling a sense of subservience or insecurity. If you look up, you can be signaling just the opposite.

The fundamental, biologically inspired point is this: adapt or die.

One must lead with creativity and follow with efficiency . . . and repeat.

In business and in life we must always be in a cycle, while constantly being aware of an equation Bob Full uses to evaluate the interplay of efficiency and creativity in his lab. "The probability of success of the project," he says, "is the value of the project divided by how long it takes. You've got to balance those things." To ensure that a high probability comes out of that equation, a simple model emerges: One must lead with creativity and follow with efficiency . . . and repeat, not the other way around and not in tandem (unless via different groups).

"Blue-sky thinking" is important at the beginning of a project, long before the deadline looms, just as gene mutations are important for a species before a life-or-death scramble for resources occurs, and just as Picasso did hundreds of studies for his masterpiece *Guernica* before ultimately painting the final canvas. If you start with efficiency, however, you confine your space of possibility before you have had the optimal amount of time to rearrange it repeatedly to reveal the best available ideas. This is why editors don't edit every sentence as a writer works, but read chapters or the whole book *after* the writer has had his or her exploratory period. (The best editors help writers find a balance between creativity and efficiency.) This is why Goethe wouldn't have burned a whole twenty years on his obsession with color if he had given himself a limit of, say, ten years to organize his thoughts on the subject, instead of the open-ended timeframe he had for his overly exploratory book.

Alternating cycles of creativity and efficiency are what the most successful living systems did (and still do), and this is the process that Silicon Valley/tech culture has effectively rebranded as their own. The key for making this work is figuring out *when* to analyze feedback from the ecology. Companies with quick feedback iterations for apps that have potential but are far from perfect . . . like the first versions of the dating app Tinder, the traffic app Waze, or the real estate app Red Fin . . . have an advantage in being able to "code-switch" rapidly from efficiency to creativity and back again, and thus best serve their product and what their consumers are telling them. They are empirically seeking the best solutions and exploring the peaks and valleys of their fitness landscape of ideas. As such, a collateral necessity of this process, according to this way of framing things, is "failure." The success of certain startups that have embraced this approach has led to all the recent hype around catchphrases like *fail forward* and *fail better*. These make for good mottos, yet their essence has been around since long before the tech community. They've discovered science.[82]

But no failure is actually taking place in such a model if done well. In science as in business—or as it *should* be in business, though the issue of revenue is always present—*failure* is when you don't learn anything, not when you disprove a hypothesis (which means you learned). So, for instance, the Silicon Valley mottos—Fail Forward is either wrong or, well, *wrong*: Either it's wrong because there is no failure in a properly designed experiment, and thus the very idea of failing is the wrong approach to the craft of science; or, Silicon Valley is wrong in its concept of failure, since to move forward in science is to learn and therefore isn't a failure at all. This view is at the heart of any and all ecologies of innovation. So maybe Silicon Valley needs some new mottos: *Learn . . . Move Forward . . . Experiment Well . . .* (not nearly as catchy).

What modern tech culture does understand very well is that the process that produces true innovation, success, and emotional fulfillment is not seamless. There are conflicts. There are delays. There are errors that have consequences, like the painful falls Ben Underwood took while he was learning how to "see" again. These disheartening bumps feel awful (and this should be acknowledged as such). But conflict can lead to positive change (if in the right ecology), since a true ecology of innovation shouldn't produce seamless results, and the brain itself again tells us why this is.

The human brain doesn't strive for perfection. We are wired to find beauty in the imperfect. Our brain is not only drawn to certainty, it is also drawn to the "noise"—the imperfections that create difference—and of course you'll remember that our five senses need contrast in order to make sense of meaningless information; otherwise it stays meaningless, and as you experienced yourself with the microsaccades self-experiment, we go blind! Take, for instance, the playing of a piano. What is the principal difference between a machine playing the piano and an expert human playing it? Why is it that the machine-based piece is aesthetically unpleasing? It sounds cold and "soulless" because—ironically—it is perfect. There are no mistakes, no hesitations, nothing that gives a sense of spontaneity and the universal human need and desire to be constantly adapting. In short, your brain evolved to find what is naturally beautiful, and what is natural is *im*perfect. This is why the process to innovate does not need to be perfect, and neither does the result.

It's how you are imperfect . . . how you are DEVIANT . . . that matters.

This spiral process of moving between creativity and efficiency is inherently imperfect because the "spaces between" are by definition zones of transition, which in biology go by the term *ecotome*—areas of transition from one kind of space to another, whether between a forest and neighboring meadow, the sea and the beach, or Neanderthals and humans. Ecotomes produce most biological innovation, but they are also where the most risk is involved, since as we all know, transitions bring uncertainty: the transition from youth to adulthood, from a familiar home to a new home, single life to married life (and often back again), childless life to parent life, working life to retired life. The key nuance is that innovation moves within the "space between." It is movement between that *is* innovation: to be innovative is not to live at the edge of chaos, but to be at the edge of chaos on "average." What this means in practice is that you must know which phase you're in at any given moment. This awareness should also extend to making sure that efficient people have been placed in an efficiency role and creative people in a creative role. If you get this wrong, you're likely to get everything else wrong in the innovation process. You have CEOs to guide the vision of a company, not COOs, whose job it is to make sure that vision gets implemented. The greatest innovators are rarely individuals, but rather groups that embody this tension between creativity and efficiency. In particular, one potentially powerful pairing inside an ecology of innovation is *the novice and the expert*.

"Undergraduates are kind of our secret weapon . . . because they don't know what can't be done," says Robert Full, referring to the "naïve" college students, sometimes young men and women not even twenty years old, who don't know what they're not supposed to believe is possible. "I would say really our major breakthroughs have been done by undergraduates. It was a sophomore to whom I said, 'We have to measure the force of an individual hair of a gecko that you can't see or hold. If we did that, we could test how it might stick.' And she took that literally!" The undergraduate found an ingenious way to actually measure the gecko hair, which Full hadn't thought could be done. "She discovered that as a sophomore. I couldn't in good conscience put a graduate student on that. She just came in and said, 'Oh! I measured it.' I said, 'What?!'" Ignorance is more than bliss—it can be achievement.

In my own lab, I tell students not to initially read in their discipline. I much prefer they just *do stuff* naïvely, to begin their own search process, which will inevitably spin out later on and require them to hunker down into a specialized efficiency mode, which their Ph.D. work will eventually require of them anyway in the end. So it's that pristinely unspecialized quality that I hope they'll preserve as long as possible, because it means they have a totally different set of assumptions than their "expert" professors (such as myself), who actually end up learning from them, as I so often do. In fact, while diversity of groups is essential for creativity within an ecology of innovation, not all diversities are the same: some kinds of diversity are better than others. In the Lab of Misfits, we typically explore a diversity of experts and naïfs (who are not ignorant, only inexperienced). This is because experts are often poor at asking good questions, because they know the questions they're not supposed to ask. Yet almost any interesting discovery comes from not asking the "right" question . . . from asking a question that seems "foolish." This is why experts can be relatively uncreative, but are instead efficient. The right kind of expert, however . . . one who embraces a way of being that is defined by play with intention (i.e., science) . . . can nonetheless recognize a good question when asked. And yet they themselves just can't ask it, which is really quite remarkable. Naïve people, in contrast, can toss out disarmingly good questions because they don't know what they're not supposed to ask. The flip side is they also don't know what constitutes a good question.

The point is that we need experts *and* novices together because they form an essential dyad. Creator types don't always know where they are going to go or what it will mean (if they are truly creators, that is, which means they are deviant). They often need a "doer" to help them, and recognize which idea is the best idea. Experts, on the other hand, are more prone to "tunnel vision," until a novice comes along and asks a question that suddenly turns the tunnel into a wide-open field. Perhaps the most celebrated example of such a novice is Einstein, who was, relatively speaking, an outsider to academia. Not professionally taught, he was naïve to the questions he wasn't supposed to ask. Of course he became *the* expert in physics, but without ever losing his naïveté. Herein lies an essential point.

It's possible to—like Einstein—maintain both expertise and naïveté inside oneself. In the Poly-PEDAL Lab, for example, where biologists work with physicists, and often neither understands the other's work very well. Full says that the people coming from different fields "are naïve *to the other's field.*" Roles constantly shift, since Full is the expert to his undergrad and grad students, but a naïf when working with, for instance, mathematicians. Recognizing people's plurality, and which role they play in any given situation, is key. "Somebody can be brilliant," Full says, "but that doesn't mean they know anything about biology." This is simply to say that for innovation we need diversity of group, which includes the "group" inside each of us. We must each foster our own internal collective. This creates an ecology of innovation with built-in contrasts. And groups bring us to leadership, which is fundamental to any ecology of innovation.

What defines a good leader? Enabling other people to step into the unseen. Our own fears about what can happen in the uncertainty of that unseen place, however, often lead us—and therefore others—astray. Our protective nature can cause us to constantly plug in metaphorical nightlights for our children, because we think we should illuminate the space for them. We need to lead kids into the darkness so that they learn to navigate it themselves.

When each of my three kids was two years old and doing what two-year-old children do, I gradually came to understand that when they were in the dark they wanted to find where the walls were—often by running full speed into them. When they went to school at Blackawton in Devon, England, Dave Strudwick, the wonderful head teacher at the time (who was also instrumental in the Blackawton Bee Project described earlier) would say to the parents, "Here, the children are allowed to climb trees" (amazing, really, that he needed to say this, since saying it explicitly implies that some parents would rather send their children to a different school than let them climb trees). Like Dave as the leader of a school, as a parent my role is not so much about turning on the lights for my children, but about letting them run in the dark (to explore), while providing walls that keep them from running too far, and letting them know with certainty that when they hit a wall and fall down they will be picked back up, since what a two-year-old wants is to both explore and be held at the same time (as does a teenager, and as does an adult . . . all of us). In this way, *they* came to define the architecture of their space of possibility (and how that space necessarily overlapped with the space of others around them . . . in this case me, their father). This requires giving children the freedom to go, and the discipline and wisdom to know when to say stop, not because their transgression triggers one of your own fears, but because they are doing something that would actually harm them. Much like the movement between creativity and efficiency, parenting is about letting children learn how to move between go and stop by building their own history of trial and error, not by adopting yours.

This means, as the world becomes increasingly connected and thus unpredictable, the concept of leadership too must change. Rather than lead from the front toward efficiency, offering *the* answers, a good leader is defined by how he or she leads others into darkness—into uncertainty. This change in leadership style is manifest in the most successful companies, the best labs, and even the best medicine.

Nick Evans is one of the United Kingdom's leading eye surgeons. Every day he works with people who are at risk of stepping into a perpetual darkness. As with any person who is suffering from an early illness, a tremendous stress arises from the uncertainty of the condition. To lead these individuals well is not to provide them with more data about their condition and symptoms—an approach called the *information deficit model*, in which the aim is to convince by providing more measured data. (This is the approach rationally used by most climate scientists—but to their detriment—when engaging in the public domain.) Rather than provide information, Nick deals directly with his patients' fear and stress arising from the uncertainty of their situation through the process of compassion—by treating them as individuals who need to understand the *why*, as well as the *how*.

A primary component of good leadership is the willingness to give . . . *unreservedly*. In the case of Bob Full, his giving drives the success of the "lab" as a whole, as well as the individuals therein, which naturally is the goal of its leader. "Almost every day I give up my best idea," says Full, meaning he regularly passes his insights about biomechanics on to his students and collaborators instead of hoarding them. "This is fine because then I select people who can take it further. When I look out I still see my curiosity finding out the answers." Sometimes these answers appear while students are still in his lab, and other times many years later. Full sees it as all the same cross-pollinating ecology, just spread out over time and space. Passions and the discoveries transcend the individuals involved.

A lab should be like family. A lab, as in any caring relationship, works well when people feel taken care of *and* safe to experiment. Trust is fundamental to leading others into the dark, since trust enables fear to be "actionable" as courage rather than actionable as anger. Since the bedrock of trust is faith that all will be OK **within uncertainty**, leaders' fundamental role is to ultimately lead themselves. Research has found that successful leaders share three behavioral traits: they lead by example, admit their mistakes, and see positive qualities in others. All three are linked to spaces of play. Leading by example creates a space that is trusted—and without trust, there is no play. Admitting mistakes is to celebrate uncertainty. Seeing qualities in others is to encourage diversity.

I would take this even further and say that the greatest leaders, whether knowingly or just intuitively, lead from an understanding of perception—knowing how to work with the human brain and its evolutionary origins, rather than against it. And of course, leaders know how to communicate this to the people they are leading. A good leader thinks in shades of gray . . . but speaks in black and white.

A good leader thinks in shades of gray . . . but speaks in black and white.

The greatest leaders possess a combination of divergent traits: they are both experts and naïve, creative and efficient, serious and playful, social and reclusive—or at the very least, they surround themselves with this dynamic. In the realm of parenting, a good parent sees the naughtiness of the behavior but the quality of the person. A good teacher creates the freedom for the student to see (as opposed to constraining them in what to see), but also sets conditions for the process of success. The greatest leaders are compassionate, are courageous, make choices, are creative, are communal—and we add the sixth fundamental quality, which is that they care about their mission; that is, they have a strong intention. Whether in a parent / child relationship, romantic relationship, or corporate relationship, creating an ecology of innovation requires knowing the why of a relationship, and even knowing the why of an individual (how they define themselves). In the case of a father and daughter or a romantic couple, that why is love (a love that they define for themselves). For a company, the why might be defined by the company's "branding DNA." This is the element that isn't questioned, while other things are. This isn't to say that it *can't* be questioned, only that to go from A to not-A, you need a purchase, a base, in addition to the ability to forgive, and thus the knowledge that things will still be OK even when you step into darkness and your brain inadvertently produces a less-than-useful response, something that inevitably happens in life. New research on the science of forgiveness has shown that it is not only good for the brain in enhancing cell growth and connectivity, but for the body as a whole.[83] A key component of practicing forgiveness is, of course, forgiving yourself when you need to. From a practical standpoint, forgiveness is forgetting an old meaning. It's the act of just stopping, of going from A to not-A . . . of "ecovi." It's why "failures" can lead to successes.

Every person embodies the dialectics / movements / tensions of perception that we have explored in this chapter. You are an expert and a novice, you are creative and efficient, you are a leader and you are led, and so contain a community within. The imperfections such a complex process creates are inevitable, along with flat-out screwups. But the deeper mistake is not trusting that your brain will reward you for stepping into the dark of uncertainty, especially when we do it alongside others . . . within the ecology of innovation. It is here and here only that we see differently and deviate, both as individuals and together. Leaders too must create *physical spaces* of deviation, since place and space are essential to human well-being. In this technical age, we seem to forget that the brain evolved in a body and the body in the world. We will never escape that essential fact, and thank God this is so!

I wrote much of this book in my home: a narrow, 200-year-old, four-story brick English townhouse in Oxford, England, that Isabel and I call The Cove. We conceive of it as a sheltered space for pirates (and non-pirates) to be different versions of themselves in all their plurality, even if our certainty-driven perception tries to convince us we're each a unitary whole. We have arranged and furnished the different spaces in our Cove to reflect the brain's movement between the two poles of creativity and efficiency, as well as the fact that spaces strongly affect our perceptions while we are occupying them. For instance, dimmer light increases creativity, whereas brighter light improves analytical thinking.[84] Ceiling height improves abstract and relational thinking, and lower ceilings do the opposite.[85] A view of generative landscapes improves generativity, whereas mild exertion temporarily improves memory and attention.[86] The Cove embodies these ideas by having each room have a different "why," which is essential for interior design and architecture. What is more, in any space it is also essential to have noise—spaces that are not predefined . . . ambiguous spaces that enable the brain to adapt. More than this, The Cove embodies an attempt to care— the desire to take active agency in creating a way of being that enables us and others to thrive *within* our misfit brains. We all must engineer our own ecology of innovation, at home and at work, both in the spaces we create and through the people with whom we populate them. Because your brain is defined by your ecology, the "personality" of the space you inhabit will necessarily shape itself accordingly.

Now that we know what constitutes an ecology of innovation, it is up to us to create one (or more than one) ourselves, wherever we see fit. Our jobs can be a lab. Our families can be a lab. Love can be a lab. Our hobbies can be a lab. Even our most banal daily routines can become a lab. These are all simply ecologies, in that what defines them is the interaction between the different parts, one of which is ourselves. Yet none of these spaces are by nature necessarily innovative, especially if they are prone to obeying the Physics of No. It is up to us to make them innovative, and by now I hope you understand why this is so essential for a fulfilled life: Doing so will enrich your brain, and new possibilities will open up for your perceptions, and this enhances your life itself.

Why Deviate?

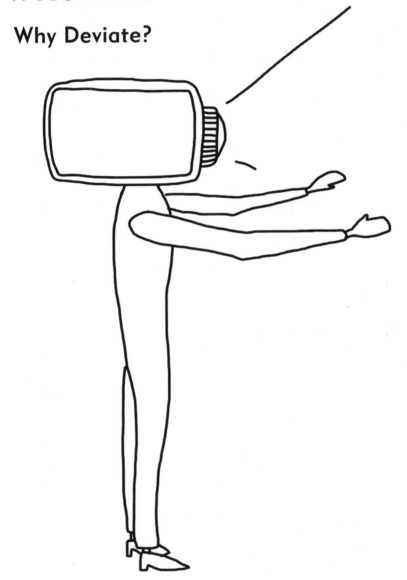

With
Deviate,
my goal was
for you to know less
at the end than you thought
you knew at the beginning. I
wanted to foster courageous doubt at
the deepest level . . . the level of your perceived
reality. By communally joining reader and author in
the same process of creating a new perceptual past, I aimed
to create new perceptions in the future. My brain now has different reflex arcs than it did when I began writing this book, and
yours should too, now that you've finished reading it (although
it's questionable whether those new arcs are the ones I intended
to create). What we both share from here forward is only the
beginning of a brain-based understanding of why we do what we
do, and why we are who we are. This shouldn't lead to a
new *fixed* way of being. The world is in flux, and what
was once useful is likely to change, maybe even
tomorrow. What I hope it does lead to,
then, is a new *plastic* way of being (like
the brain) in which you are now
always able to *see yourself*
see. Through your
own brain's
seeming-
ly

miraculous ability to let you hold multiple realities in your mind, you can reshape your perception with perception itself. By bringing informed intention to this process, you can see yourself *see yourself see differently.*

Because we now know that our perception enables us to experience life out there usefully. This is a tremendous gift earned over the cost of eons of evolution, years of development, and moments of learning through trial and—in particular—**error** (since form follows failure, not success). The embodiment of perception inside our brains in the experience of objects out there is the reason we feel we are seeing reality, although our perceptions themselves, as we know now, aren't reality. Everything you see—*everything*—exists in only one place: in here. Inside your head. Everything you experience is only taking place inside your brain *and* body, constructed in "the space between," arising from the ecology of interaction between you and your world of others, and in the space between you and the world of yourself.

It doesn't feel this way because we *project* perceptions that were created in the space between (i.e., arising from interactions between things) onto stuff out there. Thus, a red surface may appear one meter in front of you, but in fact it couldn't be closer . . . the red of that surface is *inside* you. It's as if our eyes and all our other senses combine with the rest of our brain to create a video projector. The world out there is really just our three-dimensional screen. Our receptors take the meaningless information they receive; then our brain, through interacting with the world, encodes the historical meaning of that information, and projects our subjective versions of color, shape, and distance onto things. In this sense, the ancient Greeks were close to the metaphysical truth with the emission theory of vision propounded by Plato and others, which claimed that we saw by way of streams of light that flowed out of our eyes.

Our perceptions are the feedback onto our perceptions, creating a self-reinforcing narrative, albeit an effective one for survival, and one that makes life livable. What you perceive right now is a consequence of the history of your perceptions that led to this point. And as soon as you perceive it, that perception too becomes part of your future past, thereby contributing to what you will see in the future. This is why free will lives less in the present and more in the re-meaning of past perceptions in order to change your reflexive perceptions in the future. All these meanings, including the meaning of ourselves and other people, are projected *out there*, just as we project attributes onto water, surfaces, and other objects, which we see as having different meanings in different contexts. Of course, when the other object is another human being, this process is much more complex, if not the most complex in nature. Your brain can only sense their stimuli (their sounds, their reflected light, their movement), *not* their meaning of self or their perceived meanings of you, since you can never be inside their head.

Instead, we project *everything* we perceive onto the world, including the world of others: their beauty, their emotions, their personalities, their hopes, and their fears. This isn't a philosophical position; this is a scientific explanation that I hope fundamentally changes the way you look at your thoughts and behaviors, and the thoughts and behaviors of others. While what you perceive about another person is indeed in reference to your interactions with them, your perceptions still take place internally to you, though they arise from the dialectic *space between* you and them. This means that *their* personality is in a very literal sense your personality transposed. Their fears are your fears transposed. Yes, people do exist objectively . . . just not for us, ever. Likewise, other people's perceptions of you are exactly the same. You too are created as a function of them. You contain all the personalities, fears, and colors of others you perceive.

We feel so deeply connected to other human beings—and we *are* deeply connected to other human beings—that it can feel frightening, perhaps even somehow cheapening, to know that they are just figments of our brain and body. So when I fall in love or have a profound or meaningful experience with others, is it really a *Matrix*-like dream inside my brain? In a sense, yes, but just like with a surface, with people there is an objective constancy that transcends their ambiguities that the brain must interpret.

This is their why.

In our search for certainty in an inherently uncertain life, we're often trying to find the *constant* in other people, and also in ourselves. In much the same way, we look for the invariant aspects of an object, such as its reflectance properties, which in color vision is called *color constancy*. We're effectively looking for *character constancy* in people, so that we can predict their behavior. Through prediction we feel familiarity, and in familiarity we feel secure, since to predict in evolution was to survive. In security, we are able to engage in the active vulnerability that creates trust and lets us feel *love* and *be loved*. It is possible that it is this consistency, this elusive constant that deviates from an average, that we are trying to touch in others . . . what we might call their soul. But this constant that others possess isn't a normative thing . . . it's *their personal deviation*.

The connections we feel with other people are the way our projections interact. So what we need to teach our children and each other is the ability to "just stop" when listening, not only while in conflict but always—in order to listen *differently*. **Listening is the diminishment of the answers we project onto the world. It lets in the possibility of the question, the possibility that through questions my assumptions might come to map onto your assumptions,** which is when we feel "connected." Or, alternatively, it reveals where our assumptions conflict, giving us the opportunity to be affected and enriched by spaces of possibility divergent from our own. The barrier to accepting the full humanity of others is often a lack of awareness of our own humanity, since our overriding impression is that what we see, hear, and know is the world *as it really is*. But we don't, which I hope inspires compassion. Indeed, this has been my primary motivation in writing *Deviate* . . . to inspire compassion through scientific understanding, and in doing so create the possibility of the great things that compassion (and humility), when combined with courage, creates.

By understanding how thoughts, feelings, and beliefs are intrinsically relative to one's physical, social, and cultural ecology, one can better understand the source of coherence and conflict within and between individuals. By re-seeing the process through which we are shaped by our communities, and re-meaning our historical experiences, we feel a stronger sense of skepticism and, through this, also a sense of belonging and connectedness . . . and thus courage and respect for ourselves and all things and people around us. This conception of communities encourages us to be still more humble, as it illustrates that all of us are defined by a collective ecology. So choose that ecology well, because your brain will adapt to it.

In fact, now that you know why we see what we do, to *not* enter conflict with doubt is to enter with ignorance, since intelligence is to not keep repeating the same behavior in the hope of a different outcome. At the heart of "seeing yourself see differently" is a rational reason for courage . . . the courage to occupy spaces of uncertainty. When my daughter Zanna was a child and would come to me and say she was scared to go on stage and dance, or my sons Misha and Theo go and play on a new team, or when Isabel followed bonobos in the jungles of the Congo, or my mum went to nursing school with five kids at home, or my father started his own business with little support, or my deaf great-grandmother crossed the sea to an unknown world on her own . . . my goodness, what a collection of "bad" ideas. So I completely agree with them. It *is* scary, and rightly so. Writing this book was scary, too. Doing anything that might result in utter failure is scary, and the more public that failure, the worse the fear, since we evolved a "social brain" that gives us the need to belong. Thus, to embark on what might result in failure is objectively a bad idea. When all is good around you, why on earth would you want to see what is on the other side of the hill? What an utterly bad idea, since—remember—there are more ways to die in the world than there are to survive. Yet remember those pathologically courageous fish who leave the safety of the school, risking everything but finding food?

Each of us has each kind of fish inside us. So hearing the stories of Isabel in the jungle or seeing my kids embarking on new experiences in spite of the tremendous possibility of failure . . . that *is* inspiring. We can also find tremendous inspiration in elderly people who have remained open. For a young person to do something that might fail, for a twenty-something to "fail forward" in Silicon Valley, is one thing. Relatively speaking, there's not as much to lose (though it may feel otherwise). However, when you've had the experience of life, of actual failure, of responsibility for many others, and you know the real, tangible cost of failure, but step into uncertainty anyway, that is truly amazing. That is inspiration incarnate. We all know these older people. For me, a clear personal example is a man named Yossi Vardi, who is one of the leading technology investors in Israel . . . but who is also so much more than this. He brings together the interested and interesting from around the world in his incredibly playful Kinnernet meetings, which embody the ecology of innovation that challenges assumptions, or in the school he supports for underprivileged children, which challenges the status quo of governments, and even the assumptions of age. Evidence is the fact that in his mid-seventies he is one of our favorite guests at Burning Man. And yet too often we distance ourselves from people of his age, sometimes because they reveal to us our own assumptions and biases, and even our humanity. Yet they are so often the ones who enable us to powerfully deviate, if we were to just *give in to listening*.

Research has shown that when it comes to a sense of well-being, it's difficult to beat the neurological effects of giving. This isn't to say that all behaviors aren't selfish. They are, in that they are directly (for the most part) aimed at increasing our own sense of value, which is a deep neurological need. I've argued in this book that nearly all of our perceptions, conceptions, and behaviors are in one way or another linked to uncertainty . . . either a move toward it or, more often than not, a move away from it. The deeper question, then, is how much value one adds to others in the genesis of one's own value. Might this be where we have choice? If we don't have free will in the moment . . . as what we do now is a reflex grounded in a history of meanings, this means that free will isn't about the present at all. It is about making a new future past. By choosing to re-mean the meanings of a past experience, you alter the statistics of your past meanings, which then alters *future* reflexive responses. *You* change what you're capable of.

The science of perception gives you the permission to become an observer of your own perceptions, and in doing so also gives you the need to thoughtfully deviate again and again, in order to discover the questions worth asking that might change our world . . . or might not.

Welcome to the Lab of Misfits.

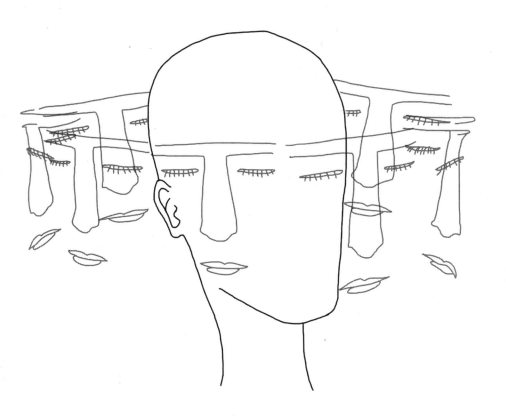

NOTES

1. Rishi Iyengar, "The Dress That Broke the Internet, and the Woman Who Started It All," *Time Magazine*, Feburary 27, 2015, accessed March 3, 2015, http://time.com/3725628/the-dress-caitlin-mcneill-post-tumblr-viral/

2. Terrence McCoy, "The Inside Story of the 'White Dress, Blue Dress' Drama That Divided a Planet," *The Washington Post*, February 27, 2015, accessed March 3, 2015, http://www.washingtonpost.com/news/morning-mix/wp/2015/02/27/the-inside-story-of-the-white-dress-blue-dress-drama-that-divided-a-nation/

3. The three articles: Karl R. Gegenfurtner et al., "The Many Colours of 'The Dress,'" *Current Biology* 25 (2015): 543–44; Rosa Lafer-Sousa et al., "Striking Individual Differences in Color Perception Uncovered by 'The Dress' Photograph," *Current Biology* 25 (2015): 545–46; Alissa D. Winkler et al., "Asymmetries in Blue–Yellow Color Perception and in the Color of 'The Dress,'" *Current Biology* 25 (2015): 547–48.

4. This and following three quotes taken from George Henry Lewes, *The Life of Goethe* (London: Smith, Elder, and Co., 1864), 98, 37, 33, 281.

5. Dennis L. Sepper, "Goethe and the Poetics of Science," *Janus Head* 8 (2005): 207–27.

6. Dennis L. Sepper, *Goethe Contra Newton: Polemics and the Project for a New Science of Color* (Cambridge: Cambridge University Press: 2003).

7. Dennis Sepper, email messages, November 11, 2014–November 14, 2014.

8. Johann Wolfgang von Goethe, *Theory of Colors* (London: John Murray, 1840), 196.

9. Johann Wolfgang von Goethe, *Faust* (New York: P. F. Collier & Son, 1909), verse 1717.

10. This and other Berkeley quotes taken from: George Berkeley, *A Treatise Concerning the Principles of Human Knowledge* (Project Gutenberg, 2003), Kindle edition.

11. A. A. Luce, *Life of George Berkeley, Bishop of Cloyne* (San Francisco: Greenwood Press, 1949), 189–90.

12. Christopher Hogg et al., "Arctic Reindeer Extend Their Visual Range into the Ultraviolet," *Journal of Experimental Biology* 214 (2011): 2014–19.

13. Tsyr-Huei Chiou et al., "Circular Polarization Vision in a Stomatopod Crustacean," *Current Biology* 18 (2008): 429–34.

14. Aquanetta Gordon, phone conversation, December 17, 2014.

15. Richard Held and Alan Hein, "Movement-Produced Stimulation in the Development of Visually Guided Behavior," *Journal of Comparative and Physiological Psychology* 56 (1953): 872–76.

16. Richard Held, Society for Neuroscience online archive, accessed on December 19, 2014, http://www.sfn.org/~/media/SfN/Documents/TheHistoryofNeuroscience/Volume%206/c5.ashx.

17. Peter König et al., "The Experience of New Sensorimotor Contingencies by Sensory Augmentation," *Conscious and Cognition* 28 (2014): 47–63.

18. Peter König, Skype conversation, December 14, 2014.

19. Nagel et al., "Beyond Sensory Substitution—Learning the Sixth Sense," *Journal of Neural Engineering* 2 (2005): 13–26.

20. Brian T. Gold et al., "Lifelong Bilingualism Maintains Neural Efficiency for Cognitive Control in Aging," *The Journal of Neuroscience* 33 (2013): 387–96.

21. Dale Purves, *Body and Brain: A Trophic Theory of Neural Connection* (Harvard College: Harvard University Press, 1988).

22. David J. Price and R. Beau Lotto, "Influences of the Thalamus on the Survival of Subplate and Cortical Plate Cells in Cultured

Embryonic Mouse Brain," *The Journal of Neuroscience* 16 (1996): 3247–55.

23. Marian C. Diamond, "Response of the Brain to Enrichment," *Anais da Academia Brasileira de Ciencias* 73, no. 2 (2001).

24. Seth D. Pollak et al., "Neurodevelopmental Effects of Early Deprivation in Post-Institutionalized Children," *Child Development* 81 (2010): 224–36.

25. Peter Hill, *Stravinsky: The Rite of Spring* (Cambridge: Cambridge University Press, 2000), 30, 93, 95.

26. Jonathan Winawar et al., "Russian Blues Reveal Effects of Language On Color Discrimination," *Proceedings of the National Academy of Sciences of the United States of America* 104 (2007): 7780-85.

27. Abby d'Arcy, "The blind breast cancer detectors," *BBC News Magazine*, Feburary 23, 2015, accessed Feburary 27, 2015, http://www.bbc.com/news/magazine-31552562.

28. Mary Hollingsworth, *Art in World History* (Florence: Giunti, 2003), 314.

29. Theresa Levitt, *The Shadow of Enlightenment: Optical and Political Transparency in France* (Oxford: Oxford Univeristy Press, 2009), 64.

30. Henry Marshall Leicester and Herbert S. Klickstein, eds., *A Source Book in Chemistry* (New York: McGraw-Hill, 1952), 287.

31. E. L. and W. J. Youmans, *The Popular Science Monthly* (New York: D. Appleton and Company, 1885), 548–52.

32. Michel Eugène Chevreul, *The Laws of Contrast of Color* (London: G. Routledge and Co., 1857).

33. https://www.youtube.com/watch?v=WlEzvdlYRes

34. Michael Brenson, "Malevich's Search for a New Reality," *The New York Times*, September 17, 1990, accessed November 14, 2014, http://www.nytimes.com/1990/09/17/arts/review-art-malevich-s-search-for-a-new-reality.html.

35. Aleksandra Shatskikh, *Black Square: Malevich and the Origin of Suprematism* (Yale: Yale University Press, 2012).

36. Herschel B. Chipp, *Theories of Modern Art: A Source Book by Artists and Critics* (Berkeley: University of California Press, 1968), 341.

37. Daniel J. Sherman and Irit Rogoff, eds., *Museum Culture: Histories, Discourses, Spectacles* (Minneapolis: University of Minnesota Press, 1994), 149.

38. Angelica B. Ortiz de Gortari and Mark D. Griffiths, "Auditory Experiences in Game Transfer Phenomena: An Empirical Self-Report Study," *International Journal of Cyber Behavior, Psychology and Learning* 4 (2014): 59–75.
39. Christian Collet et al., "Measuring Motor Imagery Using Psychometric, Behavioral, and Psychophysiological Tools," *Exercise and Sports Sciences Review* 39 (2011): 85–92.
40. Sjoerd de Vries and Theo Mulder, "Motor Imagery and Stroke Rehabilitation: A Critical Discussion," *Journal of Rehabilitation Medicine* 39 (2007): 5–13.
41. Ciro Conversano et al., "Optimism and Its Impact on Mental and Physical Well-Being," *Clinical Practice and Epidemioly in Mental Health* 6 (2010): 25–29.
42. Richard Wiseman, *The Luck Factor* (New York: Hyperion, 2003), 192.
43. Jennifer A. Silvers et al., "Age-Related Differences in Emotional Reactivity, Regulation, and Rejection Sensitivity in Adolescence," *Emotion* 12 (2012): 1235–47.
44. Cordelia Fine, *Delusions of Gender* (New York: W.W. Norton, 2010), xxiii.
45. Nicholas Epley and David Dunning, "Feeling 'Holier Than Thou': Are Self-Serving Assessments Produced by Errors in Self- Or Social Prediction?" *Journal of Personality and Social Psychology* 79 (2000): 861–75.
46. Hajo Adam and Adam D. Galinsky, "Enclothed Cognition," *Journal of Experimental Social Psychology* 48 (2012): 918–25.
47. Árni Kristjánsson and Gianluca Campana, "Where Perception Meets Memory: A Review of Repetition Priming in Visual Search Tasks," *Attention, Perception, & Psychophysics* 72 (2010): 5–18.
48. Hauke Egermann et al., "Is There an Effect of Subliminal Messages in Music on Choice Behavior?" *Journal of Articles in Support of the Null Hypothesis* 4 (2006): 29-46.
49. Michael Mendl et al., "An integrative and functional framework for the study of animal emotion and mood," *Proceedings of the Royal Society: Biological Sciences* 277 (2010): 2895–904.
50. Wolfgang Schleidt et al., "The Hawk/Goose Story: The Classical Ethological Experiments of Lorenz and Tinbergen, Revisited," *Journal of Comparative Psychology* 125 (2011): 121–33.

51. Vanessa LoBue and Judy S. DeLoache, "Detecting the Snake in the Grass: Attention to Fear Relevant Stimuli by Adults and Young Children," *Psychological Science* 19 (2008): 284–89.

52. Karen E. Adolph et al., "Fear of Heights in Infants?" *Current Directions in Psychological Science* 23 (2014): 60–66.

53. Frank J. Sulloway, *Born to Rebel* (New York: Pantheon, 1996), xiv.

54. A variety of papers by Gustavo Deco et al. have shown that resting-state network activity can be captured as attractor dynamics.

55. Steve Connor, "Intelligent People's Brains Wired Differently to Those with Fewer Intellectual Abilities," *Independent*, September 28, 2005, accessed September 29, 2015, http://www.independent.co.uk/news/science/intelligent-peoples-brains-wired-differently-to-those-with-fewer-intellectual-abilities-says-study-a6670441.html.

56. Suzana Herculano-Houzel and Roberto Lent, "Isotropic Fractionator: A Simple, Rapid Method for the Quantification of Total Cell and Neuron Numbers in the Brain," *The Journal of Neuroscience* 25 (2005): 2518–21.

57. Colin Freeman, "Did This Doctor Save Nigeria from Ebola?" *The Telegraph*, October 20, 2015, accessed November 27, 2015, http://www.telegraph.co.uk/news/worldnews/11174750/Did-this-doctor-save-Nigeria-from-Ebola.html.

58. Klucharev et al., "Reinforcement Learning Signal Predicts Social Conformity," *Neuron* 61 (2009): 140–51.

59. This claim has been made by Robert Sapolsky in several contexts.

60. Benjamin Libet et al., "Time of Conscious Intention to Act in Relation to Onset of Cerebral Activity (Readiness-Potential): The Unconscious Initiation of a Freely Voluntary Act," *Brain* 106 (1983): 623–42.

61. The four subsequent quotes come from Milan Kundera, *The Joke* (New York: HarperCollins, 1992), 31, 34, 317, 288.

62. Christian Salmon, "Milan Kundera, The Art of Fiction No. 81," *The Paris Review*, accessed August 21, 2014, http://www.theparisreview.org/interviews/2977/the-art-of-fiction-no-81-milan-kundera.

63. George Orwell, "Funny, But Not Vulgar," *Leader*, July 28, 1945.

64. This telling of the Rosetta Stone is drawn from Andrew Robinson, *Cracking the Egyptian Code: The Revolutionary Life of Jean-Francois Champollion* (New York: Oxford University Press, 2012), Kindle

edition; Simon Singh, *The Code Book* (New York: Anchor, 2000), 205–14; and Roy and Lesley Adkins, email messages, January 27, 2015.

65. Bruce Hood, *The Self Illusion* (New York: Oxford University Press, 2012), xi.

66. Robb B. Rutledge et al., "A Computational and Neural Model of Momentary Subjective Well-being," *Proceedings of the National Academy of Sciences* 111 (2014): 12252–57.

67. Ian McGregor et al., "Compensatory Conviction in the Face of Personal Uncertainty: Going to Extremes and Being Oneself," *Journal of Personality and Social Psychology* 80 (2001): 472–88.

68. Katherine W. Phillips, "How Diversity Makes Us Smarter," *Scientific American* (October 1, 2014), accessed October 20, 2015, http://www.scientificamerican.com/article/how-diversity-makes-us-smarter/

69. Frédéric C. Godart et al., "Fashion with a Foreign Flair: Professional Experiences Abroad Facilitate the Creative Innovations of Organizations," *Academy of Management Journal* 58 (2015): 195–220.

70. Paul Theroux, *The Tao of Travel: Enlightenments from Lives on the Road* (New York: Houghton Mifflin Harcourt, 2011), ix.

71. Paul Theroux, *Ghost Train to the Eastern Star* (New York: Houghton Mifflin Harcourt, 2008), 88.

72. Robert Sapolsky, *A Primate's Memoir: A Neuroscientist's Unconventional Life among the Baboons* (New York: Touchstone, 2001), 104.

73. Michael David et al., "Phasic vs. Sustained Fear in Rats and Humans: Role of the Extended Amygdala in Fear vs. Anxiety," *Neuropsychopharmacology* 35 (2010): 105–35.

74. Jennifer Lerner et al., "Fear, Anger, and Risk," *Journal of Personality and Social Psychology* 81 (2001): 146–59.

75. Rachel Nuwer, "Will Religion Ever Disappear?" BBC, December 19, 2014, accessed December 20, 2014, http://www.bbc.com/future/story/20141219-will-religion-ever-disappear.

76. John M. Gottman and Nan Silver, *The Seven Principles for Making Marriage Work* (New York: Random House, 1999).

77. Sara Lazar et al., "Mindfulness Practice Leads to Increases in Regional Brain Gray Matter Density," *Psychiatry Research: Neuroimaging* 191 (2011): 36–43.

78. M. Xue et al., "Equalizing Excitation-Inhibition Ratios across Visual Cortical Neurons," *Nature* 511 (2014): 596–600.

79. Chris Parr, "Imperial College London to 'Review Procedures' after Death of Academic," *Times Higher Education*, November 27, 2014, accessed February 28, 2014, http://www.timeshighereducation. co.uk/news/imperial-college-london-to-review-procedures-after-death-of-academic/2017188.article.

80. Harry Jerison, *Evolution of the Brain and Intelligence* (New York: Academic Press, 1973), 158.

81. Leander Kahney, *Jony Ive: The Genius behind Apple's Greatest Products* (New York: Penguin, 2013).

82. Wendy Joung et al., "Using 'War Stories' to Train for Adaptive Performance: Is It Better to Learn from Error or Success?" *Applied Psychology* 55 (2006): 282–302.

83. Megan Feldman Bettencourt, "The Science of Forgiveness," *Salon*, August 23, 2015, accessed July 24, 2015, http://www.salon. com/2015/08/24/the_science_of_forgiveness_when_you_dont_forgive_you_release_all_the_chemicals_of_the_stress_response/

84. Anna Steidle and Lioba Werth, "Freedom from Constraints: Darkness and Dim Illumination Promote Creativity," *Journal of Environmental Psychology* 35 (2013): 67–80.

85. Joan Meyers-Levy and Zhu Rui, "The Influence of Ceiling Height: The Effect of Priming on the Type of Processing That People Use," *Journal of Consumer Research* 34 (2007): 174–86.

86. Marc Roig et al., "A Single Bout of Exercise Improves Motor Memory," *PLoS One* 7 (2012): e44594.

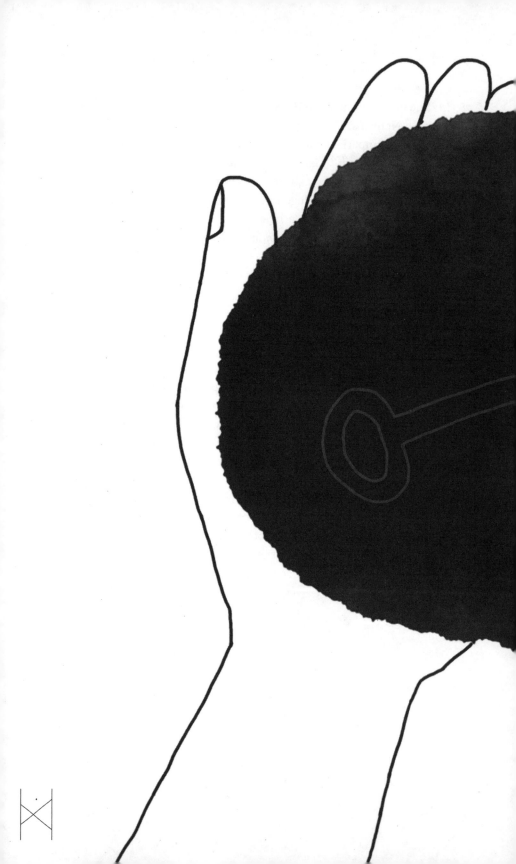

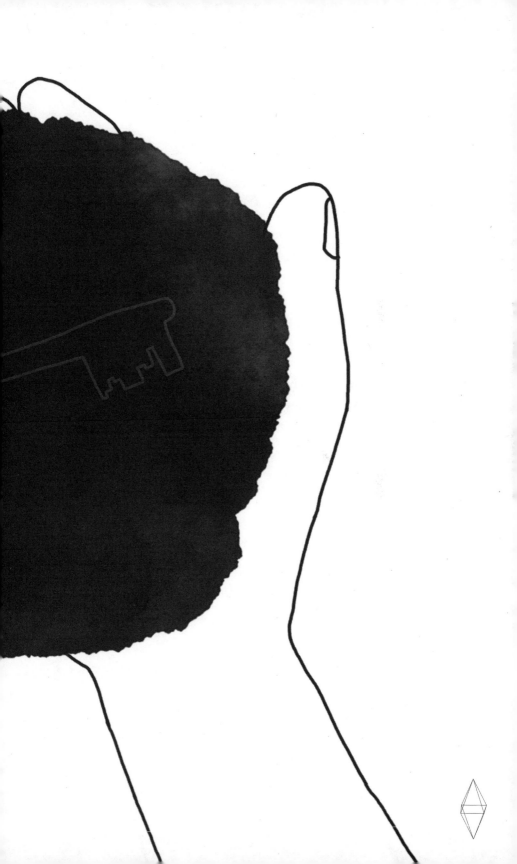

To be is to be perceived.
—George Berkeley

INDEX